W9-BQW-215

Polymer Chemistry

Polymer
Chemistry

Introduction to an
INDISPENSABLE SCIENCE

By David Teegarden

NATIONAL SCIENCE TEACHERS ASSOCIATION
Arlington, Virginia

Claire Reinburg, Director
J. Andrew Cocke, Associate Editor
Judy Cusick, Associate Editor
Betty Smith, Associate Editor

ART AND DESIGN Linda Olliver, Director
PRINTING AND PRODUCTION Catherine Lorrain-Hale, Director
  Nguyet Tran, Assistant Production Manager
  Jack Parker, Desktop Publishing Specialist

NATIONAL SCIENCE TEACHERS ASSOCIATION
Gerald F. Wheeler, Executive Director
David Beacom, Publisher

Library of Congress Cataloging-in-Publication Data

Teegarden, David M.
  Polymer chemistry : introduction to an indispensable science / by David M. Teegarden.
    p. cm.
  ISBN 0-87355-221-0
  1. Polymers. 2. Polymerization. I. Title.
  QD381.T42 2004
  547'.7—dc22
                    2004003160

Dedicated to the memory of
Katharine Morrison Teegarden
1909–2002

whose tolerance, interest in ideas, love of books,
and belief in the importance of education
have influenced all who knew her.

Contents

National Science Teachers Association

Section 4
Demonstrations and Experiments

Foreword

It is a pleasure and a privilege to introduce David Teegarden's beautiful book *Polymer Chemistry: Introduction to an Indispensable Science*. Written in a graceful style, it fills a major gap in the polymer literature. It clearly explains the connection between structure and physical properties and therefore the uses to which polymers are put in everyday life. The various types of polymerization reactions required to synthesize those polymers are also clearly explained.

David takes us through the spectacular growth of polymer chemistry over the past seven decades. These accomplishments were carried out by creative individuals who rose to the challenge and opportunity. This book will, I hope, reach younger scientists and help them to realize that they, too, can contribute.

He brings us to the present situation, where ecological concerns are increasingly important. Current research frontiers such as nanotechnology are included. Of course the recent decoding of the human Genome is ranked as a landmark of polymer chemistry, and the continuing evolution of proteomics is an exercise in structural polyamide chemistry.

David takes the reader in hand on a congenial stroll through this wonderland. He emphasizes that polymers are organic molecules differing only in size from those familiar to us in organic chemistry. He also correctly identifies viscosity, resulting from "spaghetti-like" entanglements of those long molecules, as the key difference in physical properties from those of small molecules.

David's extensive background in academia and in industrial research makes him amply qualified to present this subject. I predict that it will appeal to a wide spectrum of readers, from the technical specialist to the citizen interested in the science of the world around him or her.

H. K. Hall, Jr.
Chemistry Department
University of Arizona
Tucson, Arizona

Preface

This is a book of many subjects. Although "chemistry" is in the title, the operative word is "polymers," a subject that transcends a range of disciplines. Chemistry is included because polymers comprise molecules, many of which are synthesized by humans. As a result, polymer properties are dictated by the same laws that govern smaller, more familiar molecules. We need to understand something about the properties of these huge molecules to appreciate the extremely wide range of useful properties they possess.

Chemistry, and chemists, are only part of the story, however, as the Venn diagram on page xxiv attempts to convey. Without engineers and materials scientists, we would have interesting molecules but few useful materials. Finally, we need the skills of design engineers, architects, fiber specialists, packaging engineers, and a host of other product designers to figure out how to produce the tremendous assortment of objects fabricated with polymers. The knowledge and creativity of people in these fields and others combines to provide the materials that define our way of life and that we tend to take for granted. Although we have a long way to go, scientists and manufacturers are making significant progress in producing these materials more efficiently from natural resources, improving our ability to reuse and recycle them, and lessening the impact on important environmental issues such as ozone depletion and global warming.

Ironically, the field of polymers, so central to each of our lives, is woefully neglected in our science curricula at all levels. High school and college chemistry textbooks often tuck a token section or chapter near the back. Rarely are polymeric molecules presented as examples during the coverage of the usual topics. Most students, science majors or not, have an extremely poor understanding of the many essential ways polymers affect their lives.

The primary purpose of this book is to provide a resource that helps teachers introduce polymer concepts in their classes. Basic principles are developed and compared to "small-molecule" fundamentals where possible. The subject matter of this book assumes some background in general and organic chemistry, although much of the coverage is quite general and descriptive. Chapter 4 provides a general history of polymers and could be used in a number of different classes. Chapter 9 discusses the disposal and recycling of polymers, topics that fit nicely in environmental science curricula. Elizabeth Dabrowski's insightful "About This Book" provides a teacher's perspective on how to use the text.

We have been living in the Polymer Age for quite some time. My fervent belief, shared by my many colleagues, is that we must increase everyone's awareness of these essential materials, not just that of the chemists and engineers who will be working with them professionally. I hope this book will facilitate that educational process in some small way.

Industrial workers who are beginning to work with polymers should also find this book a useful introduction to the subject. Laboratory technicians, technical staff, and managers in many organizations can learn the vocabulary quickly and begin to gain an appreciation for the field of polymers and its many facets.

Acknowledgments

The author expresses his warm thanks to Wayne T. Ferrar, Adam Freeman, Christine Landry-Coltrain, and Dennis J. Massa, all stimulating industrial research colleagues for many years. Each is a patient teacher with the gift of explaining difficult concepts in understandable terms. Henry K. Hall, Jr., Emeritus Professor of Chemistry at The University of Arizona, has provided the author valuable assistance, insight, and encouragement for more decades than either of us wishes to acknowledge. His many scientific contributions have helped shape the Polymer Age. His love of polymers and passion for teaching have influenced generations of students. Lynn M. O'Brien, Professor of Chemistry at Nazareth College of Rochester, provided many helpful suggestions on the treatment on natural polymers. All of these individuals read the manuscript carefully, offering valuable suggestions and critical comments that have resulted in a more accurate and readable text. Their enthusiasm for the project has been very encouraging.

James Shannon, Mendon High School, Pittsford, New York, and Aileen Svereika are gifted and dedicated high school chemistry teachers who offered many helpful suggestions. Debbie Liana, SUNY-Buffalo, and Paul Wesson, Northwestern University, provided valuable comments from the students' perspective. I thank all of these people for their enthusiasm, skill, and willingness to somehow squeeze this task into their extremely busy schedules.

I am grateful for the encouragement and support of Ed Schofield and the management of the Research Laboratories at Eastman Kodak Company. I also thank Ellen Dietterick for her valuable assistance with literature searches. In addition, I am grateful to Stephen Teegarden for his assistance in drawing some of the figures.

Betty Smith and Claire Reinburg of the NSTA Press have been extremely helpful and just terrific to work with. Thank you for supporting me in this project and for your patience!

Finally, I express a special gratitude to my wife, Carole. Her many comments on the manuscript were extremely beneficial and helped clarify a number of confusing passages. In addition, her love, patience, and encouragement during the manuscript preparation were invaluable.

David Teegarden
January 2004

About This Book

How often do your students ask you, "What does this have to do with real life?" Probably every day in every class. Students voice this concern so often that the school at which I teach made "Connecting Learning to Life" a school improvement goal. Similarly, with this book, you have in your hands a way to answer that question and achieve that connection by saying, "It is all about polymers, and polymers are all around us today."

Polymer chemistry is definitely a growth industry, but most chemistry teachers have had few polymer chemistry or materials science courses even at the college level. This textbook gives a teacher a thorough introduction to the chemistry of polymers, both synthetic and natural. Moreover, it is easy to understand and enjoyable to read.

I would like to suggest ways in which you can use this book in the classroom. An obvious way is that you can use this textbook as just that—a textbook in an elective polymer chemistry course. You could us it as a textbook in an advanced placement chemistry course during those weeks between the AP examination and the end of school.

Polymer of the Week

But there are also ways you can use this textbook to introduce students to the fascinating subject of polymers in a regular high school chemistry course. Some books on the market encourage teachers to teach their classes about a different element, compound, or demonstration each day or week. Why not apply that idea to polymers? Each week choose a different polymer, such as nylon or acrylic or DNA (deoxyribonucleic acid). Each day devote about five minutes of class time to the "Polymer of the Week." Here's how you then could proceed:

- On Monday, have students discuss where they might find the polymer of the week. Is it found in nature or what consumer products or medical products are made with that polymer?
- On Tuesday, introduce a discussion on the type of polymer—elastomer, chain-growth, step-growth, fiber, or copolymer—and introduce some of the actual chemistry of polymers.
- On Wednesday, use Chapter 4 and its excellent history of polymers to discuss the development of the polymer of the week.
- On Thursday, discuss how the products that were mentioned in the class discussion on Monday are produced, such as the forming of bottles through blow molding.
- On Friday, use Chapters 1 and 9 for a risk-benefit analysis of the Polymer of the Week.

Using Each Chapter

Yet another way to use this textbook is to use polymers as the examples for discussions or explanation of concepts covered in a chemistry course. Chapter 1 can be used in conjunction with the first chapter of any textbook where a discussion of "What is chemistry?" occurs. A teacher wants students to know the importance of chemistry in their everyday lives and the good things that have been the result of chemistry, but it is also necessary to discuss some of the problems that have arisen because of the careless use of chemistry. (You can also refer to Chapter 9, "Disposal, Degradation, And Recycling; Bioplastics," for another angle on responsible chemistry.)

With Chapter 2, "What Are They?" you can use a long list of covalent molecules to explain covalent bonding in the context of something as ubiquitous as a plastic grocery bag. You can also use polymer reactions as examples of composition reactions.

Across the Disciplines

Chapter 3, "Natural Polymers," might not find use in the regular curriculum of a high school-level chemistry course, but would be excellent in a biology course at any level. The explanation of natural polymers is very clear and introduces the chemistry of these biomolecules in an easy-to-understand fashion that could be used in even an introductory high school biology course.

Chapter 4, "The History of Polymers," offers a chance to do an interdisciplinary activity with a social studies teacher. Students can learn how the development of polymers is an integral part of the economic development of the twentieth century. Students can gain an appreciation of how conflicts were often the reason for the development of synthetic polymers to replace natural ones or ones that required monomers that were no longer accessible. In the chemistry course, students can learn more about the companies discussed in the history of polymers and what products they market in the twenty-first century.

Elementary and Advanced Chemistry

Chapter 5, " Polymer Synthesis," is probably most useful in a second-year chemistry course that touches on the topics of stereochemistry and some organic chemistry.

Density, a topic that is taught in the early weeks of a chemistry course, is introduced in Chapter 6, "Polymer Solutions and Dispersions." Why not use some plastic samples for density experiments? Have the students identify the plastic from its density. This will enforce the concept that density is an intensive property of matter. The discussion of viscosity can also fit into a presentation on H-bonding. The explanation of polymer viscosity is very useful for a teacher using the "Chemistry in the Community" curriculum. It is an excellent reference for the experiment that has the students determining viscosity by timing the rate of fall of a plastic bead in various organic liquids.

Chapter 7, "Physical Properties," provides examples of intermolecular forces and how they affect the state of matter and the physical properties of familiar compounds.

Chapter 8, "Polymer Processing," can be introduced on those days when you don't have enough time to begin a new concept before a vacation or a free day. If you gather some photographs or computer images of the machines mentioned in this chapter, students can learn about the manufacturing of chemicals. Those photographs are also a good start for a discussion of the differences and similarities of careers as chemists vis-a-vis chemical engineers or materials scientists. They are also useful for a discussion of the development of a new "product" from the research laboratory bench to the pilot plant to the actual manufacturing plant.

Chapter 9, "Disposal, Degradation, and Recycling," can be used for a class project for Earth Day. Students have sent projects on the effects of the environment on plastics out into space on the International Space Station. Classes can do a similar project on a smaller scale by studying the effects of the weather on some plastic samples kept outside the classroom window.

Writing across the curriculum is being strongly encouraged in many schools. Chapter 10, "A Glimpse of Things to Come," can be the source of ideas for a student essay. Ask students to write about the future. Ask them to suggest new uses for current polymers or uses of existing polymers that might require the invention of totally new polymers. These ideas do not have to be currently feasible. This could be a bit of science fiction writing, but it could also encourage a great deal of creative thinking from students. These essays should find their way out onto a bulletin board so that other students and teachers could see how learning is being connected to life in the chemistry classroom.

Experiments

If you have been highlighting a polymer a week, the first four experiments in Section 4—"Free Radical Polymerization," "Synthesis of Nylon," "Synthesis of Polyesters in the Melt," and "Synthesis of a Polyurethane Foam"—are excellent demonstrations to intersperse with the content as it is presented. If you want your students to actually perform the experiments, it might be best to wait until the end of a first-year chemistry course when the students have developed their laboratory techniques to the greatest extent. Another use for the four experiments would be to introduce a different one each quarter and discuss the polymer produced in the experiment. This is a good way to use the information on polymer chemistry if time does not permit the presentation of a Polymer of the Week.

One of these four experiments can find a special use. "Step-Growth Polymerization: Synthesis of a Polyurethane Foam" can be a special part of a study of stoichiometry. Although it does not deal with molar ratios, it does ask the students to calculate how much of each reactant is needed to produce an object of a desired volume. This reviews concepts such as unit conversions and volume.

You can use "Polymer Precipitation" at the beginning of the school year as a safety demonstration. Compare what happens when an egg is cooked to what happens when a concentrated acid is added to a raw egg. The students will observe a very similar denaturing of the protein. This can be a lesson on why students must wear safety goggles in the laboratory and can also be used during a presentation on the dangers of acids and bases.

"Gels from Alginic Acid Salts" teaches two areas of chemistry simultaneously. In addition to introducing polymer chemistry to a class, it also shows periodic trends. The experiment looks at the reactions of a sodium alginate solution in the presence of divalent metallic cations. Ions that are suggested for use include those of calcium, magnesium, iron, cobalt, nickel, copper, and zinc. Students can compare main group ions and transition metal ions. Students can also compare a divalent ion to a trivalent ion by observing the reactions with magnesium and aluminum ions. Students can compare the trends in reactions down a group on the periodic table by observing the results of the reactions with calcium and magnesium. Several of the period four transition metals are used in the experiment, and students can observe the colors of the beads formed based on the different ions used to form the beads. The experiment also introduces some coordination chemistry, especially in the reaction of the copper bead with ammonia solution.

Density is always introduced sometime in the early weeks of the first-year chemistry course. Traditional laboratory experiments have students determining the density of water or ethanol. Teachers usually set up a density column with various liquids and solids to demonstrate differences in density. Why not substitute "Densities" for this experiment? It has directions on setting up a "polymer density column," and the students can use part one of the experiment to understand the concept of density. This can be a second density experiment after the traditional experiment. The students can be presented with the solutions of various densities and several samples of known polymers and an unknown polymer. (Some scientific supply companies sell polymer samples to be used for specific gravity experiments. Used in density experiments, they eliminate the problems of floating or air bubbles.) Ask students how density would be used to sort the polymer samples. This can be a more open-ended experiment and may prove somewhat challenging but would be excellent, especially for honors students.

"Experiments with Films" is an application of some practical chemistry. While it involves some very simple techniques of stretching and tearing, it illustrates a very critical aspect of the testing of polymers in physical testing laboratories. The fabric used in the escape slides of airplanes is tested by stretching it taut and then repeatedly hitting it with a pointed metal probe. What is the point of this test? Airline passengers are told to take off shoes, jewelry, and even glasses if they must use the escape slide but often they do not. Women have been known to keep on high-heeled shoes. The puncture test is a way of seeing if the polymer can withstand a pair of high heels. Students can discuss the direction of the lines in a garbage bag and why they are oriented in that direction. (Yard waste can puncture the bag, but when you lift it up, you don't want it to tear and leave the bottom of the bag of garbage on the floor or sidewalk.) Use this experiment to illustrate the difference in strength between covalent bonds and intermolecular forces. If students pull along the transverse direction (samples marked TD or B), some covalent bonds are actually being broken, and this increases the difficulty in tearing the sample. If students pull along the machine direction (samples marked MD or A), intermolecular forces are merely being overcome. The experiment recommends looking through the polymer at a light source as it is being stretched. If possible, use a set of polarizing filters to look at the polymer as it is being stretched and the areas of strain and stress are colorfully evident. Containers from cassette tapes or adhesive tape can be observed between polarizing filters. The colors observed show the areas of strain put on the plastics during the forming of the object. These ideas can even be introduced in a study of spectra during a presentation on the atomic theory to show how light can be used to determine the structure of materials. A more sophisticated version of "Experiments with Films" could use a strain gauge probe for the calculator or computer so that the actual force needed to tear the sample could be determined. This would give the students a greater appreciation of actual techniques used in the research laboratory.

These are just a few of the methods you can use with this text. As you continue working with the book, you will discover more ways to help your students learn about the fascinating and pervasive subject of polymers.

Elizabeth Dabrowski
Magnificat High School, Rocky River, Ohio

About the Author

David Teegarden is a research scientist with Eastman Kodak Company in Rochester, New York. He earned an A.B. in chemistry at Ohio Wesleyan University and an M.S. and Ph.D. in organic chemistry at the University of Michigan. He spent 17 years as a college professor, teaching primarily courses in organic and polymer chemistry. He has spent more than 18 years in industrial research and development, working on the synthesis of specialty polymers for various imaging applications.

Teegarden makes frequent presentations to science teachers' workshops and conventions, including the National Science Teachers Association national convention, and has written extensively for polymer chemistry journals and other science publications. He is involved with the American Chemical Society Polymeric Materials Science and Engineering Division, the Science Teachers Association of New York State, and NSTA.

Notes on Equations and Units

Chemical Equations Involving Polymers

The observant reader will notice that some of the equations in this book are not balanced. Although unbalanced chemical equations can be distracting, there are times when trying to balance them is even more distracting. Chemical equations involving polymers certainly can be balanced, but doing so often leads to confusion. For example, consider this simple reaction from Chapter 2 indicating the formation of nylon-6:

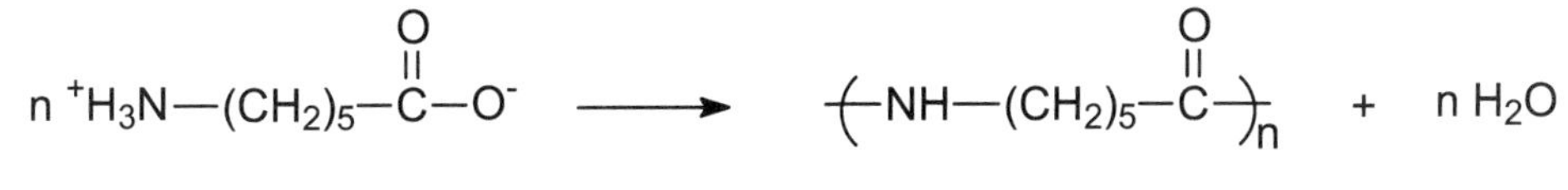

$$n \; {}^{+}H_3N-(CH_2)_5-\overset{\overset{\displaystyle O}{\|}}{C}-O^{-} \longrightarrow \left(\!-NH-(CH_2)_5-\overset{\overset{\displaystyle O}{\|}}{C}\!-\right)_{\!n} + \; n \; H_2O$$

Here we have attempted to balance the equation using n to indicate some number of moles. Unfortunately, this symbolism leads to confusion, since the subscript n does not stand for the total number of moles of product, but in fact is the average number of moles of amino acid units in a polymer molecule. Thus in situations like this we will avoid the confusion by leaving out the prefixes and understand that normal stoichiometry applies.

A Note on Units

We will use SI units throughout this text as much as possible. When another system of measurement is much more common, however, we will use the more conventional or convenient unit.

Physical data that appear in this text are taken from reliable sources, but it is not unusual to find different values in different sources for the same polymer. For example, the glass transition of polyethylene can be found listed as -128°C, -80°C, and -30°C (almost a 100°C range!), supposedly for the same material. Discrepancies such as this can arise from differences in structure between samples, either because the materials are inherently different, or because of the way the samples were prepared. In some cases, different measurement techniques give different values for the same sample. Although we will try to choose values that are representative, do not be surprised to find somewhat different values in any given source.

Some Useful Conversions:

mass

1 kg	=	2.2 lb
1 tonne (metric)	=	1×10^3 kg
	=	1.10 ton (U.S.)

length

1 mm	=	1000 micrometer (μm)
1 mm	=	1000 nm
1 mil	=	0.001 in = 12.7 μm

area

1 mi^2	=	640 acres (U.S.)
	=	259 hectares
	=	2.59 km^2

pressure (force per unit area)

1 psi	=	6.895×10^3 Pascals (Pa)

Polymer Science Relationships

Venn diagram illustrating the relationships of the field of polymer science to the core physical sciences, as well as to the biological and materials sciences, all feeding technology that provides products and materials to society.

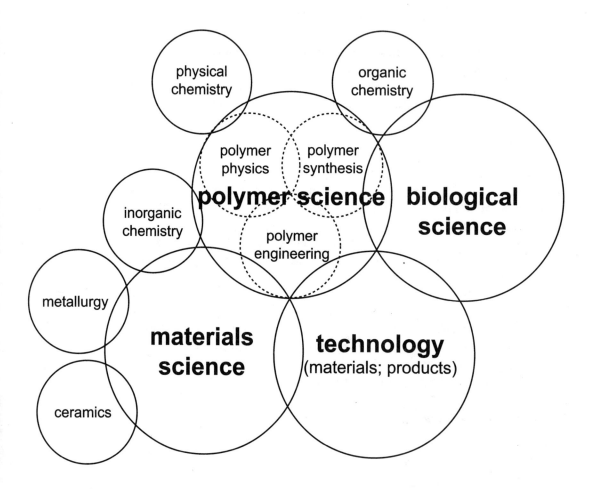

National Science Teachers Association

Section *1* Introduction

Chapter *1*

Chemistry is the mother of all technologies.
E. A. Rietman

They're Everywhere!

Major Types of Polymers

The Increasing Importance of Synthetic Polymers

"Just One Word"

Less Mass, Yet More

There Is No Free Lunch
References Cited
Other Reading

As we begin to study polymers and their properties, let's consider why they literally surround us. Why are they so important and so pervasive? Why are so many of the things we buy, eat, wear, consume, and discard polymers? As we look around, we see very few objects that do not contain polymers. In fact, our bodies are mostly polymeric: our bones, muscle, DNA, enzymes, skin, and hair, to name just a few parts. Much of what we eat, too, is polymeric or contains polymers: meat (protein), potatoes and pasta (starch), milk (protein), and green vegetables (cellulose). Many of our processed foods have been modified by adding polymers—for example, instant soups, ice cream, milk shakes, cheese, sausage, jams and jellies, and whipped "cream." Consider the materials that package our foods, to protect them physically, to keep water out (or in), to protect them from microorganisms—most of these materials are polymers. Our clothes are all made up of polymers, either natural (cotton, linen, wool, silk) or synthetic (polyester, nylon, rayon). We depend upon many synthetic polymers for medical applications, including sterile packaging, surgical garments, blood bags, drug delivery devices, artificial organs and joints, replacement blood vessels, and skin grafts for burn patients.

Our houses are built of wood (cellulose and lignin), sheathed with particleboard (wood chips pressed with plastic resin), wrapped with plastic sheeting, clad with siding (vinyl), and decorated with plastic shutters. Buildings with wood siding are covered with paint. Inside, water flows through plastic pipes, the floors are covered with tiles (vinyl) or carpeting (polyester, nylon), the walls are covered with wallpaper (vinyl) or painted (acrylics), and the bathtubs, shower stalls, kitchen counters, and sinks are fabricated from synthetic polymers (acrylics and polyesters). For the past 50 years or so, our automobiles have contained increasing amounts of polymers, primarily to reduce mass. Originally, plastic parts replaced metal in and around the dashboard and in trim pieces. Seat coil springs were replaced with foam (polyurethane). Under the hood, polymers became increasingly common in heating ducts, fan housings, and electrical and electronic components. Outside, fenders, grilles, lamp housings, and wheel covers are no longer made of metal. An increasing number of automobile bodies are now molded from polymers rather than stamped from metal. Toys are made of tough plastics; boats, personal watercraft, and snowmobiles are molded with fiberglass-filled resins; downhill skis, bobsleds, and bulletproof vests contain extremely strong synthetic polymer fibers.

When we shop, we often pay with "plastic." We talk with our friends on cell phones, listen to CDs on our personal stereos, watch TV, and surf the Net on our computers. All of these devices consist mostly of plastics. The list truly seems endless.

Major Types of Polymers

Let's start by developing an overview of the major types of polymers. We can categorize polymers in a number of ways. We will develop chemical as well as structural classifications later in the text when we learn about their synthesis and properties. However, to begin, we will divide them on the basis of origin and function. We have already alluded to two different types: natural and synthetic. Table 1-1 lists several types of *natural* polymers and provides examples of each. As their name implies, natural polymers occur in nature.

Table 1-1. The major types of natural polymers.

Type	Examples
polysaccharides	
structural	cellulose (wood, cotton, flax, hemp)
reserve	amyloses, amylopectins (starches [potato, corn, tapioca], glycogen, dextrans)
gel-forming	gums, mucopolysaccharides
proteins	egg white, gelatin, enzymes, muscle, collagen, elastin, silk, wool
polynucleotides	DNA, RNA
polyisoprenes	natural rubber, gutta-percha, chicle
polyesters	poly(3-hydroxybutyrate), cork
lignins	binder for cellulose fibers, cell walls

Also referred to as biopolymers, they are synthesized in the cells of all organisms. It is interesting to note that two of the most prevalent types, polysaccharides and proteins, each contain diverse compounds with extremely different properties, structures, and uses. For example, the protein in egg white (albumin) serves a much different function (nutrition) from that in silk or wool (structural). Likewise, the properties of starch and cellulose could hardly be more different. Although each is made up of polymers based on the condensation of glucose, the final molecular structures differ dramatically. Both sustain life, but in completely different ways. We will discuss natural polymers in more detail in Chapter 3.

However interesting natural polymers are, the focus of this book will be on synthetic or man-made polymers. The reasons for this emphasis will become obvious as we increase our understanding of polymer science. For now, however, let the following two statements suffice:

■ Although nature provides us with a huge variety of materials for specific uses, it doesn't give us polymers that can be easily molded (i.e. are plastics).

■ Natural polymers are very difficult to synthesize in the laboratory.

In Chapter 4, we will discover that the field of polymer science essentially began when scientists chemically modified natural polymers to prepare new materials with improved properties. Some of these reactions are still commercially important today. However, the specificity in the structures of most biopolymers themselves makes their laboratory synthesis extremely difficult.

So what do we mean by *synthetic* polymers? Simply put, these are compounds that are synthesized or made by man using known chemical reactions and processes. The distinction for polymers is exactly the same as that between natural and synthetic small-molecule organic compounds. In fact, most synthetic polymers are made from organic starting materials and are themselves (very large) organic compounds. Almost all of the starting materials come from petroleum. Inorganic polymers (e.g., silicones) are both known and important, but are far fewer in number compared to their organic cousins. We will explore the properties of silicone rubbers in Chapters 5 and 7.

Consider Figure 1-1, which shows the conversion of crude oil into petrochemicals and fuel. The process of producing organic compounds from petroleum is called refining. Less than 10% of petroleum that is refined is converted into petrochemicals, almost all of it being converted into gasoline, diesel or jet fuel, or heating oils. Petrochemicals are converted into a variety of organic compounds, including solvents, organic compounds such as dyes,

Figure 1-1. The small fraction of petroleum that is converted to polymers.

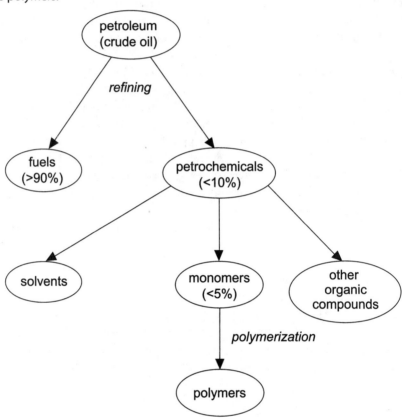

fertilizers, drugs, and food additives, and into monomers, the starting materials for making polymers.

Table 1-2 separates synthetic polymers into broad categories that describe their ultimate function or property. We are familiar with some of these terms.

■ *Plastics* are materials that can be heated and then molded or shaped into useful objects. More specifically, *thermoplastics* can be processed more than once by reheating, while *thermoset* resins react chemically during initial heating and molding to form a permanent network. The term *plastic* is often loosely used to describe all polymers, sometimes in an unfavorable way. As we can see in Table 1-2, however, several other categories of polymers exist, each with its own set of properties.

■ *Fibers* are polymers with high strength in one dimension that can be processed into long strands. The range of strength for fibers is truly impressive, from those used for textiles (clothing, carpets; relatively weak), to commercial monofilament fishing line up to 130 km long (80 land miles!), to materials that can be woven into bulletproof vests (aramids).

Table 1-2. Examples of synthetic polymers.

Type	Examples	Typical Uses
plastics	polystyrene, poly (methyl methacrylate), poly(vinyl chloride)	bottles, toys, automobiles, seemingly everything
fibers	nylon, polyesters, polyaramids	clothing, disposable diapers, tennis racquets, carpets, fishing line
films	polyethylene, polyesters	packaging, grocery and garbage bags, paints, photographic films
elastomers	polybutadiene, polyisoprene	tires, golf balls, condoms, latex gloves, rubber bands
adhesives	epoxies, poly(vinyl alcohol), polycyanoacrylates	white glue, epoxy cement, "instant" glue

- *Films* have two-dimensional strength and can be processed into flexible thin sheets. Grocery, garbage, and dry-cleaning bags, food wrap, and photographic film are common examples. *Coatings* (e.g., paints) can be included in the category of films that adhere to the surface of other materials.
- *Elastomers* are rubbery polymers with varying ability to stretch.
- *Adhesives* include glues and sealants and may be rubbery or hard after drying or setting.

Although this list covers the majority of polymer types, additional classes of polymers exist, including some specialized ones that we will take up later in the text. Also, some polymers fall into more than one category depending upon application. For example, some of the polymers formulated into paint have excellent adhesive properties, for obvious reasons.

The Increasing Importance of Synthetic Polymers

We understand something of the importance of natural polymers. Cellulose is relatively stiff and allows trees and plants to grow toward the sun. Starch is an efficient molecule for storing energy. Proteins such as enzymes catalyze biological reactions. DNA (deoxyribonucleic acid) and RNA (ribonucleic acid) are responsible for the storage and transmission of genetic information. But why do we find so many synthetic polymers in our lives?

The first reason is the most obvious: *replacement*. Because most synthetic polymers are organic, their densities are quite low compared to metals and ceramics. Consider an object as simple as a drinking tumbler to be used by a major airline.

- Common glass has a density of approximately 2.6 g/cm^3, while the density of polystyrene used by the airline is 1.0. Also, the polystyrene tumbler can be less than one-half as thick as the glass because the polymer is stronger and less brittle. Therefore, the plastic tumbler has a mass only about 10% to 20% of the mass of the glass one. Advantage number one.

■ Assuming the item could be molded with an inexpensive plastic (polystyrene is a cheap commodity plastic), the cost for the plastic tumbler would be significantly lower. And for about the same price, it will also have the airline's symbol printed on the side. Advantage number two.

■ Someone is going to drop a tumbler from time to time. The plastic one is much less likely to break, and, if it does, it will not shatter. Advantage number three.

■ The tumbler made out of glass is expensive enough that it will have to be returned, washed, and reused. The cheap plastic one will simply be thrown out. Advantage number four.

You can see that using polystyrene tumblers saves the airlines time and money. And that is why only first-class passengers receive their beverages in glass tumblers!

Repeat this scenario with just about any common object you can think of, and you begin to understand why polymers are so common. Plastics come in an extremely wide range of compositions, with different physical and mechanical properties, chemical resistance, and cost. They can be molded, blow-molded, extruded, and shaped into an infinite variety of one-, two-, and three-dimensional shapes and objects. They can be made optically clear, translucent, opaque, textured, and multicolored. They have made available to all of us household items that were once enjoyed only by the extremely wealthy (Wascher 1988). Thus the evolution of what many call the *Plastic Age* was inevitable.

The packaging industry provides many excellent examples for the substitution of plastics for older, traditional materials. For example, plastic beverage containers have become increasingly common, replacing aluminum, steel, glass, and paper for many types of drinks. Mass is one factor, obviously. In addition, as shown in Table 1-3a, the cost of the energy required to produce a kilogram of each material varies dramatically. These are average numbers for generating *virgin* (new) material and do not take into account recycling.

Table 1-3b takes the message one step further. What is the energy cost for producing a given container? Again, because the density of most plastics is so low, the low bulk energy cost (Table 1-3a), coupled with the low mass of the container, results in a per container cost at least one-tenth that for aluminum or glass. How could food and beverage manufacturers ignore such dramatic economic figures?

Much of the food we buy, either in the grocery store or in fast-food restaurants, is wrapped in some form of plastic. Some of this is for convenience and marketing (e.g., trendy prepackaged snacks with separate compartments and wrappings for each type of food), while some packaging is used to retard spoilage. Obviously, decisions about what foods to encase in plastic, and what plastics to use, depend upon economic, marketing, and technical (polymer science and plastics engineering) considerations.

Table 1-3. Energy requirements for beverage packaging material (adapted from Guillet 1997, courtesy of Wiley-VCH).

1-3a. Thermal energy required to produce 1 kg of material.

Material	(J/kg) (x10^{-5})*
aluminum	26.7
steel	5.0
glass	2.9
paper	2.5
plastic	1.1

1-3b. Energy per container

Container	Volume	Mass (g)	Energy/Container (J) (x10^{-6})*
aluminum Pepsi can	12 oz (355 mL)	14.0	3.7
returnable glass beer bottle	12 oz (355 mL)	238.8	6.8
paper milk carton	1 pint/16 oz (473 mL)	26.1	0.66
plastic milk bottle (HD polyethylene)	1 pint/16 oz (473 mL)	30.9	0.34
plastic Grower's Pride OJ (PET)	1 pint/16 oz (473 mL)	45.6	0.50
plastic Pepsi bottle (PET)	20 oz (591 mL)	27.4	0.30

* J = joule

"Just One Word"

Mr. McGuire: *I just want to say one word to you. Just one word.*
Ben: *Yes, sir.*
Mr. McGuire: *Are you listening?*
Ben: *Yes, sir, I am.*
Mr. McGuire: *Plastics!*

(*The Graduate*, 1967)

This scene was particularly funny in 1967 when Mr. McGuire advised recent college graduate Ben to pursue a career in plastics. A then-young Dustin Hoffman as Ben tried to answer politely, but the dismay and bewilderment of his reaction made this scene one of the most memorable in film history. (If you aren't familiar with this classic movie, talk to someone older—they will be.) The replacement of conventional materials with polymers that began in earnest during and following World War II initially resulted in plastic products notably inferior to the quality of the originals. The terms *plastic* and *synthetic* took on the meaning of cheap and impermanent imitations. Being part of an industry that seemed only to turn out an increasing number of such items was not what Ben or many other college graduates dreamed of.

But the replacements have continued unabated ever since, and advances in polymer science coupled with improvements in processing soon brought about higher quality and more sophisticated products. By 1976, plastics surpassed steel to become the most consumed synthetic material in the United States (Wascher 1988). New polymers, as well as new uses for old polymers, continue to find niches in consumer applications with ever-increasing volume. Not surprisingly, the use of plastics has exploded since their first introduction in the midtwentieth century. This is shown for plastics in Figure 1-2. The reasons behind the dramatic growth of this new industry and the ways it has changed our lives will be the underlying theme of this book.

Figure 1-2. Worldwide consumption of plastics from 1950 to 2010 (Szabo 2002, reprinted courtesy of Rapra Technology, Ltd.).

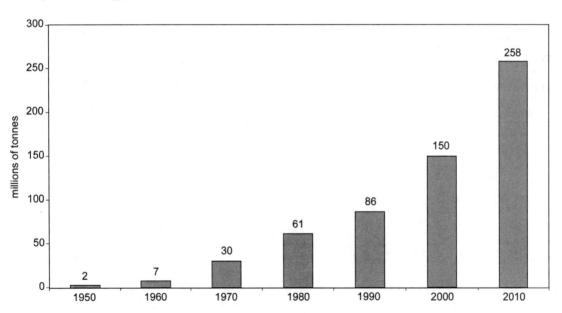

Sometimes the replacement of a traditional item with a simple plastic one can provide a huge improvement in quality of life. In many parts of Africa, people must walk long distances to obtain water. Traditionally this job fell to the women, who carried the water on their heads in ceramic vats. Children couldn't carry a heavy, breakable vat, which could be a family's most valuable possession. Many families could not afford more than one.

Enter the plastic bucket. Light, inexpensive, and not very fragile, it enabled the children to assume some responsibility for obtaining water, thus freeing women for other tasks (Fenichell 1996).

Less Mass, Yet More

If the simple replacement of "conventional" materials with polymers constituted the whole story, a study of polymers would be far less interesting than it actually is. The second major reason for the prevalence of so many polymers in our society results from the possibility of *unique properties*. The study of polymers, polymer science, encompasses several disciplines, including chemistry, physics, and engineering. The range of properties one can obtain with macromolecules is extremely broad and truly fascinating. In addition, we see an ever-increasing demand for more sophisticated materials. Thus, industrial corporations as well as government agencies have long funded research in the synthesis and processing of new polymers, resulting in a wide range of commercial products with properties that are not possible with other materials. Such materials have resulted in the development of entirely new technologies, such as space travel, integrated circuits, computers, optical fiber, water-based paints, artificial organs, and automobile tires that last 60,000 miles or more.

Polymers can be made that are electrically insulating, semiconducting, conducting, or magnetic. When subjected to electrical fields, exposed to light or changes in temperature, some polymers change color or switch from transparent to opaque (e.g., liquid crystals). Some polymers change shape when subjected to an input of energy, while others cause solutions to become more or less viscous or to form a gel when it is desirable to do so. Many materials do not consist solely of polymers, but are *composites* containing substances called *fillers*. Some fillers are inexpensive inorganic materials such as glass, minerals, or carbon black. Alternatively, fibrous or platelike particles can be used as fillers to reinforce a polymer, making the composite material stronger than the polymer alone. Examples of *reinforcing agents* include glass fiber, carbon fiber, and other polymers (e.g., aramid or nylon fibers). We will discuss some of the technological applications of composites in more detail in Chapters 8 and 10.

How far we have come since the period of cheap imitations! Consider for a moment the lowly automobile bumper. Until fairly recently, bumpers were constructed of steel and electroplated with chromium to make them shiny. Even minor accidents usually necessitated the replacement of a steel bumper, whose job was to protect the automobile body from damage. Now, most cars have plastic bumpers that offer a number of advantages. Not only are they much lighter, they are also very tough, meaning they are able to withstand significantly greater impact before undergoing damage. In addition, they can be molded in complex shapes so they become integrated into the overall design of the automobile. The plastic can be colored, either the same as the rest of the car or as a complement. The colorant is not just on the surface, so minor scrapes and bumps are hardly noticeable. And finally, they do not rust.

There Is No Free Lunch

Not surprisingly, in addition to the many advantages of polymers comes the inevitability of some disadvantages. The plastic tumbler discussed above doesn't have the "heft and feel" of a glass one. It isn't as classy. And significantly, the disposal of an increasing number of these and many other plastic items causes sometimes alarming increases in the volume of municipal waste entering our landfills. Extremely few of these materials are biodegradable, and many cannot be directly reused.

So how do we balance the increasing pressure to introduce even more plastic objects into our lives while minimizing the impact on our environment? This is an extremely important question, one that we will address when we discuss issues surrounding the recycling and disposal of polymers in Chapter 9.

Meanwhile, let's continue to build our understanding of polymers. So far, we have talked briefly *about* polymers and why they are so prevalent in our lives. We have distinguished natural polymers from synthetic ones and have classified the latter based on properties. We have also suggested the promise of making materials with very interesting and unusual properties. In the next chapter we will explore exactly what polymers are and will probe the origin of their unusual properties. We will begin to understand polymers as chemical molecules similar to, but very different from, small organic molecules.

References Cited

Fenichell, S. 1996. *Plastic: The making of a synthetic century.* 9-10. New York, NY: HarperBusiness.

Guillet, J. 1997. Environmental aspects of photodegradable plastics. *Macromolecular Symposia*, 123: 209–24.

Szabo, F. 2002. The world plastic industry. *International Polymer Science and Technology 29(6)*: T71–T77.

Wascher, U. S. 1988. GE plastics materials strategy. *Polymer News* 13: 194–98.

Other Reading

Elias, H-G. 1987. *Mega molecules.* New York, NY: Springer-Verlag.

Emsley, J. 1998. *Molecules at an exhibition: Portraits of intriguing materials in everyday life.* Gallery 5. New York, NY: Oxford University Press.

Seymour, R. B., and C. E. Carraher. 1990. *Giant molecules: Essential materials for everyday living and problem solving.* New York, NY: Wiley Interscience.

Chapter 2

What Are They?

The term *polymer* derives from *poly*, the Greek prefix for "many," and *meros*, Greek for "part." We can think of a *mer* as being the smallest unit making up a polymeric molecule, or *macromolecule* (*macro*, Greek for "large"). So in its simplest form, a polymer results from the buildup of many single units, or *monomers*, forming long chains. Each molecule is one of these long chains. This can be easily visualized in a number of ways. Snap beads or beads on a string, for instance, can be assembled in long strands to represent polymer chains. This is shown in Figure 2-1a, in which little circles are linked together to form a linear chain. In general, for a macromolecule to have useful physical properties, the chains must contain hundreds to thousands of monomers. Compounds with far fewer monomer units in a chain than this are termed *oligomers* (*oligo*, Greek for "few"). There are no strict definitions distinguishing a (high) polymer from an oligomer. The chemical process of monomers reacting with each other to form polymeric molecules is called *polymerization*.

The macromolecule depicted as a cartoon in Figure 2-1a is called a *homopolymer*, because all of the monomers in the chain are the same. If the monomer were styrene, the homopolymer would be named polystyrene. Note that in Figure 2-1b, however, there are two different types of circles making up the chain, each type representing a different monomer. This represents a *copolymer*. Now things begin to get more interesting! In bonding together two different monomers into a long chain, one can quickly think of several arrangements or sequences that might result. Three distinct relative distribu-

Figure 2-1. Representation of different types of polymer chains. Filled circles represent one monomer, open circles another.

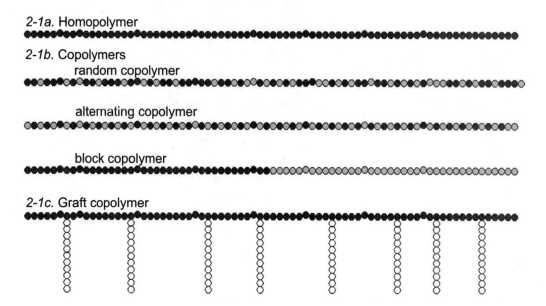

2-1a. Homopolymer

2-1b. Copolymers
random copolymer

alternating copolymer

block copolymer

2-1c. Graft copolymer

tions include *random*, perfectly *alternating*, and *block*, all of which are depicted in the figure. Each of these three copolymers can have significantly different properties from the other two. The challenge, then, is for the synthetic polymer chemist, the person who carries out the polymerizations, to design chemical reactions that will cause the two monomers to react with each other exactly as desired.

Polymerization of three monomers produces a terpolymer, four a quaterpolymer, and so on. For simplicity, the term copolymer is often used to describe any polymer except a homopolymer.

Not all polymer chains are linear. One could envision *branched* chains, in which the main chain contains some number of side chains or branches. The branches can contain the same monomers as the main chain, or the monomers in the branches can be different as illustrated in Figure 2-1c. The chemical process of attaching side chains to existing linear chains is called grafting. Because *graft copolymers* can have long segments of one monomer bonded to long segments of another monomer, their structures can loosely resemble that of block copolymers. Other chain architectures are possible and known, such as star-shaped molecules or highly branched molecules that resemble trees. Thus the properties of polymers are dependent upon not only the monomer composition but also the structure or architecture of the molecules. We will explore the major chemical reactions that can be used to synthesize polymers in Chapter 5.

How Big Are They?

As we mentioned above, macromolecular chains need to be quite long for the polymer to achieve desirable mechanical properties. How do we express this? Consider a homopolymer of styrene (phenylethene, C_8H_8), depicted in the structure below:

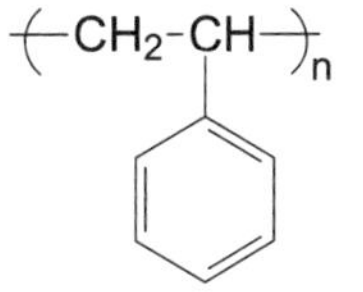

$$\left(CH_2\text{-}CH\right)_n$$

The structure inside the parentheses (in this case a styrene unit) is the *repeat unit* of the polymer, indicating that every styrene unit is bonded on either side with another styrene unit, resulting in a chain. Groups on the ends of the chain are generally not shown, although they can usually be inferred from the method used to carry out the polymerization. The letter *n* on the outside of the parentheses represents the average number of styrene units bonded together in the homopolymer and is called the *degree of polymerization* (DP). As we will learn in Chapter 5 when we discuss the chemistry of polymerizations, not all polymer chains in a macromolecular sample contain the same number of monomer units. Therefore, *n* represents an average for the distribution of chain lengths.

Say we have a polystyrene sample with an average DP of 50. Because we know the formula mass (FM) for styrene (104), we can calculate an average molar mass for the polymer:

molar mass = DP x FM = 50 x 104 g/mol = 5200 g/mol

This is called the *number-average molar mass* (M_n) and is the average molar mass of each chain. The chains have a chemical group attached at each end, the nature of which depends upon how the polymer was synthesized. For high polymers the mass of these end groups is negligible, however, and is ignored.

Other ways of expressing polymer molar mass exist, the most common of these being the *weight-average molar mass* (M_w). M_w is always higher than M_n. The ratio M_w/M_n is called the *molar mass distribution* or the *polydispersity*. The larger this ratio, the greater the distribution of chain lengths in the sample. Figure 2-2 plots the molar mass curves for two different polymers, one with a narrow polydispersity and one with a relatively large polydispersity. As we will see in later chapters, many polymer properties are dependent upon both the molar mass and the polydispersity.

Figure 2-2. Molar mass data and plots for a) a polymer with low polydispersity and b) a polymer with normal polydispersity. M_n and M_w are indicated for polymer b. (Data courtesy of Thomas Mourey, Eastman Kodak Company)

Polymer	M_n	M_w	Polydispersity (M_w/M_n)
a	7,090	7,270	1.03
b	130,000	305,000	2.35

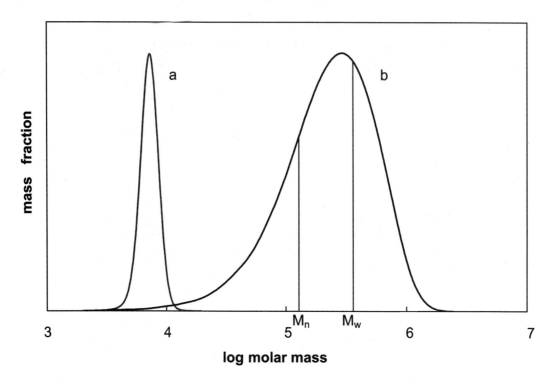

As we mentioned at the beginning of this chapter, polymer chains need to be quite long for the polymer to have useful mechanical properties. Now we can address what we mean by long. The polystyrene example above with a DP of 50 is actually considered to have a rather low molar mass, although we would probably not consider it to be an oligomer (which would have a DP more like 5 to 10). A high polymer of styrene would have a DP in excess of 100 (M_n 10,000) or more. Let's try to understand why.

How Long Is Long Enough?

Earlier we developed a model for a polymer chain by using a long strand of beads. We could also choose other one-dimensional chain models, such as spaghetti noodles or strands of Easter basket grass. Based on their diameters, neither of these objects is nearly long enough to be a really accurate model, but either suffices nicely here. Cooked spaghetti noodles, as well as the grass strips, even though they are linear, do not tend to line up well when mixed with other "chains." The noodles or strands mix randomly, their various chains becoming entangled. Picture trying to remove one spaghetti noodle from a bowl of cooked noodles, or one strand from a bag of Easter basket grass. As you move one of these "chains," other chains around them are also disturbed. Because the neighbors inhibit the motion of a given chain, it is not easily separated from the mass.

And so we arrive at the basis for the strength of many polymers, especially plastics—*molecular entanglement!* Picture hitting a glass windowpane with a hammer. The same blow to an equivalently shaped piece of Lexan (polycarbonate) or Lucite (poly[methyl methacrylate]), both transparent plastics, does not break the polymer. In fact, the plastic "glass" is somewhat resilient, causing the hammer to bounce off of it.

For polymers, the chains must reach a critical length before effective entanglement can occur. This number varies from one type of polymer to another. For poly(methyl methacrylate), for example, the critical entanglement degree of polymerization is approximately 300 (M_n approximately 30,000).

Polymers—Molecules with Unique Properties

How different polymers are from small-molecule organic compounds! The latter exist as gases, liquids, or solids. Polymers often behave as solids, although some are viscous liquids or gums. When heated, most organic compounds boil or sublime. Most polymers do not. Polymers do not exist as gases. The molecules in pure organic compounds are all exactly alike. Molecules in a sample that have different structures from the given compound are called impurities. In a polymer sample, because there is a molar mass distribution, it is likely that very few molecules are exactly the same length! The molecules in organic liquids are in random motion and easily pass by one another. Macromolecules are entangled, severely restricting molecular motion. Polymeric liquids are very "thick" or viscous.

Nonetheless, the same rules that govern the interactions of small organic molecules are in play as well with macromolecules (or any molecules). The normal forces of attraction and repulsion between atoms and molecules do not discriminate on the basis of molar mass or chemical composition.

It might be instructive to consider the impact of molar mass on the physical properties of a regular series of saturated organic hydrocarbons, called a homologous series. Let's start with the simplest member, methane, CH_4, and, on paper, sequentially add one CH_2 unit (or *methylene group*) to build the series. To start, consider writing methane as CH_3-H, and then insert CH_2 into the bond between the CH_3 and the H. The result is ethane, CH_3-CH_2-

H, or CH_3-CH_3. Continuing this process builds the series. This series is shown in Table 2-1, along with the name, boiling point, and melting point for each compound. Compounds with from 1 to 4 carbon atoms are gases at room temperature, the boiling point increasing significantly as each additional methylene group is added. The compounds with 5 to 15 or so carbons are liquids under atmospheric conditions. The lower-boiling compounds are useful as nonpolar solvents (e.g., paint thinner).

Table 2-1. From small molecules to polymers: the effect of chain length on physical properties.

# C's	Structure	Name	Boiling Point (°C)	Melting Point (°C)
1	CH_3-H (CH_4)	methane	-164	-182
2	CH_3-CH_3	ethane	-89	-183
3	CH_3-CH_2-CH_3	propane	-42	-190
4	CH_3-CH_2-CH_2-CH_3	butane	0	-138
5	CH_3-$(CH_2)_3$-CH_3	pentane	36	-130
6	CH_3-$(CH_2)_4$-CH_3	hexane	69	-95
7	CH_3-$(CH_2)_5$-CH_3	heptane	98	-91
8	CH_3-$(CH_2)_6$-CH_3	octane	126	-57
9	CH_3-$(CH_2)_7$-CH_3	nonane	151	-51
10	CH_3-$(CH_2)_8$-CH_3	decane	174	-30
12	CH_3-$(CH_2)_{10}$-CH_3	dodecane	216	-10
16	CH_3-$(CH_2)_{14}$-CH_3	hexadecane	288	20
20	CH_3-$(CH_2)_{18}$-CH_3	eicosane	343	37
30	CH_3-$(CH_2)_{28}$-CH_3	triacontane	—	65
50	CH_3-$(CH_2)_{48}$-CH_3	pentacontane	—	92
100	CH_3-$(CH_2)_{98}$-CH_3	hectane	—	115
>>100	CH_3-$(CH_2)_n$-CH_3	polyethylene (HDPE)	—	138

As the chain lengths grow, the number of van der Waals interactions between segments of the molecule increases. This causes the boiling points as well as the melting points to increase. At some point (16 to 17 carbon atoms), the melting point rises above room temperature, and the subsequent compounds become waxy solids. At 100 carbons, the melting point is 115°C. If we continue increasing the number of carbon atoms until we achieve a high polymer, the limiting melting point of 138°C is reached. At this point, the crystals (called *crystallites*) have reached optimum size and order for the polymer. Increasing the number of carbon atoms beyond this point (making the chains even longer) will not cause any further increase in melting temperature. The resulting polymer is known as polyethylene, used for grocery bags, milk bottles, and a number of other applications.

How Are They Made and How Can We Name Them?

We added one methylene group at a time to build the homologous series above. Yet the name of the final compound was based on a two-carbon monomer. Why not call the polymer polymethylene instead of polyethylene? The answer is that the most common nomenclature scheme uses the starting *monomers* to identify the polymer (a *source-based* system). Thus one usually polymerizes ethylene to make what constitutes on paper a long chain of methylene groups bonded together. However, we would draw the structure of the polymer as repeating units consisting of two carbon atoms to indicate that the polymer was prepared from a two-carbon monomer. This is illustrated in the following equation:

Equation 1

ethylene monomer   polyethylene   polymethylene

This type of polymerization was traditionally called *addition polymerization*, because each monomer has at least one double bond to which are added elements of two other monomers. The monomer and the repeat unit in the polymer have the same empirical formula. A more precise term for this type of reaction is *chain-reaction* or *chain-growth polymerization*, because this describes the chemistry by which the polymerization takes place. Many of our most familiar polymers are prepared using this chemistry, which will be developed in Chapter 5.

To name a copolymer prepared by chain-growth polymerization, we simply link the names of the two (or more) monomers with *co* and precede the names with *poly*. The set of names is usually placed in parentheses to avoid confusion.

Table 2-2 lists the common names and structures of a few chain-growth polymers. For each of these the common name of the homopolymer and its repeat unit are also shown. Two examples of copolymers are also provided. The subscripts m and n indicate only the total number of repeat units of each monomer, not the arrangement in the copolymer. In other words, these are presumed to be random, not block, copolymers. All of the monomers in Table 2, a and b, polymerize by addition to a carbon-carbon double bond by a chain-growth mechanism.

Table 2-2. Names and structures for common monomers and polymers.

Monomer	Structure	Polymer Name	Polymer Structure
a. Chain-growth homopolymers			
acrylamide	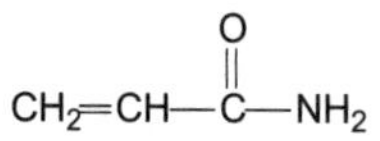 $CH_2\!=\!CH\!-\!\overset{\overset{O}{\|\|}}{C}\!-\!NH_2$	polyacrylamide	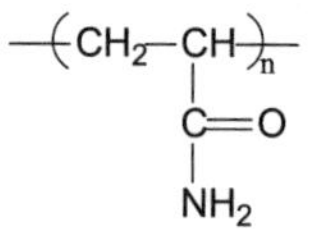 $-\!(CH_2\!-\!CH)_n\!-$ with $C\!=\!O$, NH_2
butyl acrylate	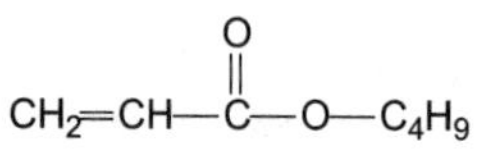 $CH_2\!=\!CH\!-\!\overset{\overset{O}{\|\|}}{C}\!-\!O\!-\!C_4H_9$	poly(butyl acrylate)	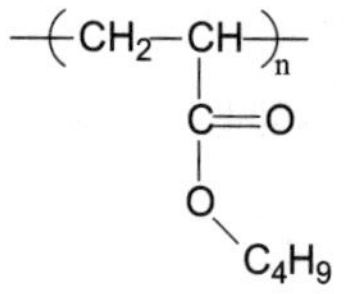 $-\!(CH_2\!-\!CH)_n\!-$ with $C\!=\!O$, O, C_4H_9
methacrylic acid	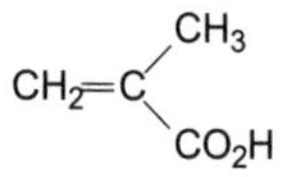 $CH_2\!=\!C\!\overset{CH_3}{\underset{CO_2H}{\big<}}$	poly(methacrylic acid)	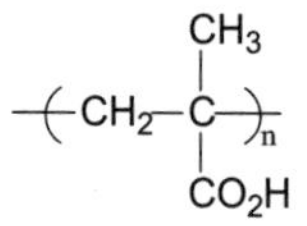 $-\!(CH_2\!-\!\overset{CH_3}{\underset{CO_2H}{C}})_n\!-$
methyl methacrylate	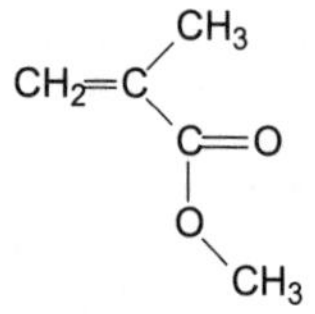 $CH_2\!=\!C\!\overset{CH_3}{\underset{C=O,\ O,\ CH_3}{\big<}}$	poly(methyl methacrylate)	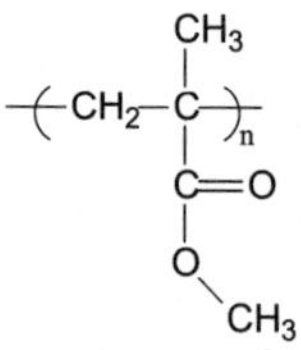 $-\!(CH_2\!-\!\overset{CH_3}{\underset{C=O,\ O,\ CH_3}{C}})_n\!-$
α-methyl styrene	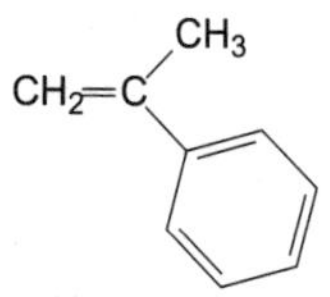 $CH_2\!=\!C\!\overset{CH_3}{\underset{C_6H_5}{\big<}}$	poly(α-methyl styrene)	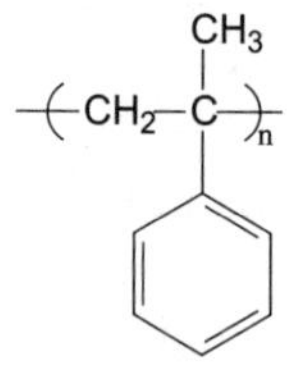 $-\!(CH_2\!-\!\overset{CH_3}{\underset{C_6H_5}{C}})_n\!-$
tetrafluoro-ethylene	$CF_2\!=\!CF_2$	polytetrafluoro-ethylene	$-\!(CF_2\!-\!CF_2)_n\!-$

Table 2-2, continued

Monomer	Structure	Polymer Name	Polymer Structure
vinyl chloride	$CH_2{=}CHCl$	poly(vinyl chloride)	$-(CH_2{-}CH)_n-$ with Cl
isoprene	$CH_2{=}C{-}CH{=}CH_2$ with CH_3	polyisoprene	$-(CH_2{-}C{=}CH{-}CH_2)_n-$ with CH_3 (1 structure)

b. Chain-growth copolymers

Monomer	Structure	Polymer Name	Polymer Structure
styrene and acrylonitrile	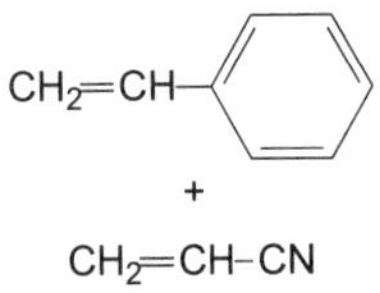 $CH_2{=}CH{-}$ (phenyl) $+$ $CH_2{=}CH{-}CN$	poly(styrene-co-acrylonitrile)	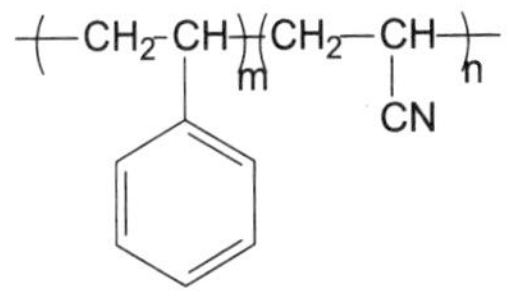 $-(CH_2{-}CH)_m(CH_2{-}CH)_n-$ with phenyl and CN
ethylene and vinyl acetate	$CH_2{=}CH_2$ $+$ $CH_2{=}CH{-}O_2CCH_3$	poly(ethylene-co-vinyl acetate)	$-(CH_2{-}CH_2)_m(CH_2{-}CH)_n-$ with O_2CCH_3

c. Ring-opening homopolymers

Monomer	Structure	Polymer Name	Polymer Structure
ethylene oxide	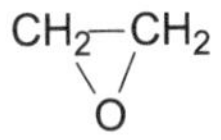 $CH_2{-}CH_2$ with O (epoxide)	poly(ethylene oxide)	$-(CH_2{-}CH_2{-}O)_n-$
caprolactam	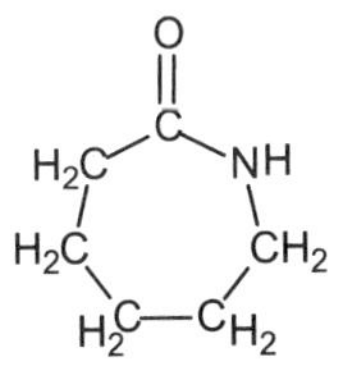 caprolactam ring: $H_2C{-}C({=}O){-}NH$, H_2C, CH_2, $H_2C{-}CH_2$	polycaprolactam (nylon-6)	$-(NH{-}CH_2CH_2CH_2CH_2CH_2C({=}O))_n-$

Certain cyclic compounds undergo polymerization reactions called *ring-opening polymerization*. Two examples are included in Table 2-2, c.

Besides chain-growth and ring-opening polymerization, another very common type of polymerization is called *step-growth polymerization*. Many, but not all, of the monomers that polymerize by this mechanism undergo condensation reactions. That is, two different functional groups react to form a new functional group while evolving some small molecule such as water, methanol, or HCl. An older term for such polymerizations is *condensation polymerization*. An example is given below in which adipic acid (1,6-hexanedioic acid) reacts with hexamethylenediamine (1,6-diaminohexane) to produce poly(hexamethylene adipamide) and H_2O. This polymer is an example of a polyamide, or a nylon. This polymer is also called nylon-6,6, because both the diacid monomer and the diamine monomer provide 6 carbon atoms in the polymer chain.

Equation 2

$$HO-\overset{O}{\underset{\|}{C}}-(CH_2)_4-\overset{O}{\underset{\|}{C}}-OH \ + \ H_2N-(CH_2)_6-NH_2 \longrightarrow \left(\overset{O}{\underset{\|}{C}}-(CH_2)_4-\overset{O}{\underset{\|}{C}}-NH-(CH_2)_6-NH\right)_n \ + \ H_2O$$

Note that in this type of polymer, the repeat units have different empirical formulas from those of the starting materials. Further, the repeat unit in the nylon example shown consists of two monomer fragments. Although the presence of two different monomer fragments in chain-growth polymerization would constitute a copolymer, that is not the case in the nylon example. Both fragments are used to make up the amide bond, and the resulting polymer is a homopolymer. Clearly we can make copolymers of step-growth polymers by, for example, reacting one diacid with two different diamines.

Nylon-6,6 was prepared from a diamine and a diacid. Another way to make a polyamide, at least on paper, is to begin with one monomer that contains both an amino group and a carboxylic acid group (an amino acid), as shown below for 6-aminohexanoic acid:

Equation 3

$$^+H_3N-(CH_2)_5-\overset{O}{\underset{\|}{C}}-O^- \longrightarrow \left(NH-(CH_2)_5-\overset{O}{\underset{\|}{C}}\right)_n \ + \ H_2O$$

The polymer would be named either poly(6-aminohexanoic acid) or nylon-6. Note that the repeat unit in this polymer comes from only one monomer, and is therefore different from the repeat unit for nylon-6,6.

Note also that nylon-6 was listed in Table 2-2 as the product of the ring-opening polymerization of the cyclic monomer caprolactam. So here is an example of the same polymer that can be made from (at least) two different monomers by two different mechanisms. This illustrates the complexity of our classification and nomenclature schemes, because polycaprolactam, poly(6-aminohexanoic acid), and nylon-6 all specify the same polymer. This is summarized in the reaction scheme below, which shows that 6-aminohexanoic acid can be polymerized by a step-growth polymerization to nylon-6, or converted to the alternate monomer caprolactam. Caprolactam can then be polymerized in a separate step to the nylon. Although all of this might appear somewhat confusing and redundant, it illustrates that different routes to the same polymer provide the polymer chemist with important options. As we will see in Chapter 5, different synthetic reactions can lead to subtle but significant differences in the final polymer.

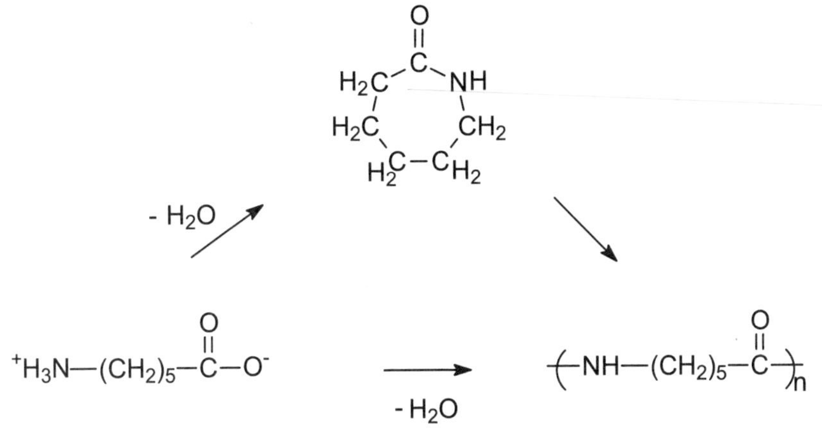

Table 2-3 lists examples of monomers and their corresponding polymers, with names, for a number of step-growth polymers.

Summing Up

Macromolecules can be made as homopolymers or copolymers, and usually have straight or branched chains. One way to make branched chains is by graft polymerization. The monomers in copolymers can exist in different sequences, such as random, alternating, and block. The number of monomer units in a chain is called the degree of polymerization, from which one can calculate an average molar mass. Long chain length (high DP or molar mass) gives rise to molecular entanglement, the primary phenomenon that distinguishes polymers from small molecules. Many polymers derive their inherent mechanical strength from the entanglement of their macromolecular chains.

Although additional reaction mechanisms will be covered in Chapter 5, three important reactions for preparing polymers are *chain-growth*, *ring-opening*, and *step-growth* polymerization. Chain-growth and ring-opening polymers are usually named by placing *poly* before the monomer name(s). Step-growth polymers are named by placing *poly* before the chemical name of the repeat unit of the polymer. We should point out here that the International

Table 2-3. Names and structures for common monomers and step-growth polymers.

Monomer 1	Structure	Monomer 2	Structure
a. Polyesters			
terephthalic acid	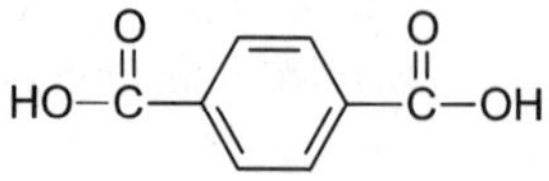	ethylene glycol	$HO-CH_2CH_2-OH$
isophthalic acid	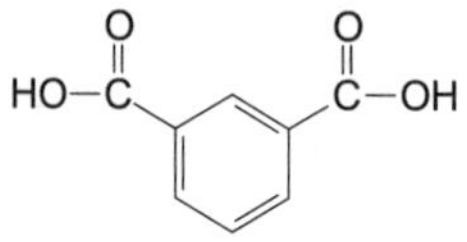	1,4-butanediol	$HO-CH_2CH_2CH_2CH_2-OH$
b. Polyamides			
6-amino-hexanoic acid		none	
adipic acid		hexamethyl-enediamine	$H_2N-CH_2CH_2CH_2CH_2CH_2CH_2-NH_2$
terephthalic acid	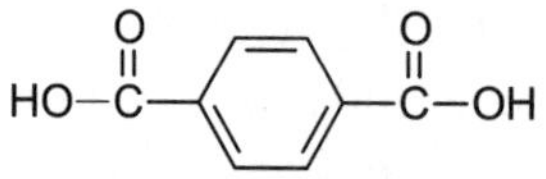	phenylene-diamine	$H_2N-(CH_2)_6-NH_2$

Polymer Name	Polymer Structure
poly(ethylene terephthalate)	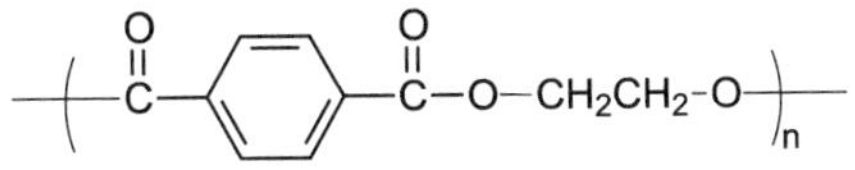
poly(butylene isophthalate)	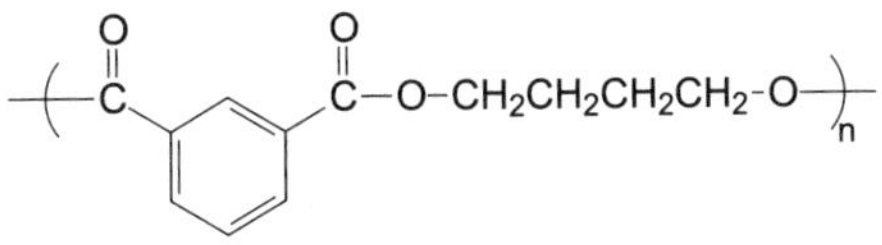
poly(6-amino-hexanoic acid) [nylon-6]	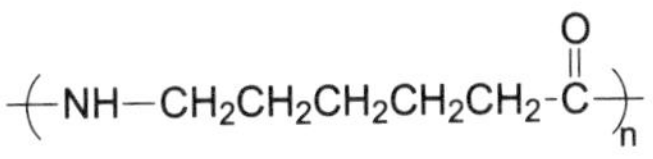
poly(hexa-methylene adipamide) [nylon-6,6]	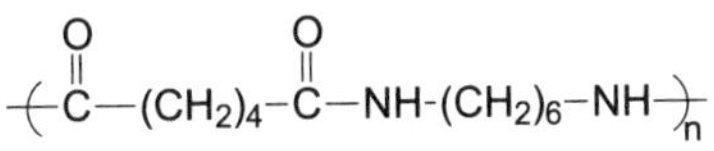
poly(phenylene terephthal -amide)	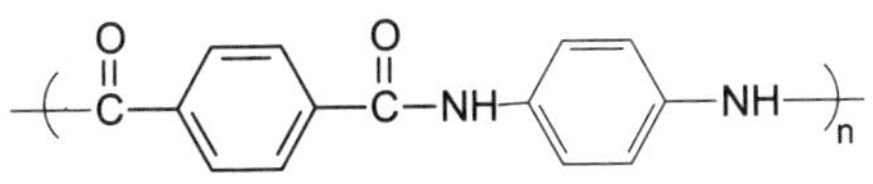

Union of Pure and Applied Chemists (IUPAC) has developed systematic nomenclature rules for polymers. As is the case with many small-molecule organic compounds, the IUPAC names are often complex and cumbersome. Therefore, polymer scientists often use common or abbreviated names, sometimes even product names. Poly[1-(methoxycarbonyl)-1-methylethylene] (IUPAC) is almost always referred to as poly(methyl methacrylate), or PMMA, or even Lucite. IUPAC discourages the use of trademarked names, however, unless it is important to refer to a specific commercial product. The IUPAC Macromolecular Nomenclature Commission recognizes a number of trivial names for common polymers (Metanomski 1999).

References Cited

Metanomski, W. V. 1999. Nomenclature. In *Polymer handbook*, eds. J. Brandrup, E. H. Immergut, and E. A. Grulke, 4th ed., I/1–I/12. New York: John Wiley and Sons.

Other Reading

Campbell, I. M. 2000. *Introduction to synthetic polymers*. 2nd ed. New York, NY: Oxford University Press.

Elias, H-G. 1997. *An introduction to polymer science*. New York, NY: Wiley-VCH.

Nicholson, J. W. 1997. *The chemistry of polymers*. 2nd ed. London: The Royal Society of Chemistry.

Stevens, M. P. 1999. *Polymer chemistry: An introduction*. 3rd ed. New York, NY: Oxford University Press.

Walker, F. H. 2001. Fundamentals of polymer chemistry: I. *Journal of Coatings Technology* 73 (912): 75–79.

Walker, F. H. 2001. Fundamentals of polymer chemistry: II. *Journal of Coatings Technology* 73 (913): 125–30.

Without natural polymers, there is no life.
Dr. Hans Uwe Schenck

Natural Polymers

Polymers that exist in nature, called *biopolymers,* include a large and diverse range of compounds. Table 1-1 lists the major classes of natural polymers. In this chapter we will discuss the most important types—their chemical makeup, key properties, and where they are found. We will focus more on the chemical and physical properties of natural polymers and less on their biological synthesis or physiological function. The references at the end of the chapter provide additional information.

To start, let's divide natural polymers into two major categories. *Homologous biopolymers* consist of only one type of monomer unit—for example, proteins (amino acid units). *Heterologous biopolymers,* as their name implies, contain more than one class of monomer units. An example would be glycoproteins, which contain both carbohydrate and protein portions. Heterologous polymers are often block or graft copolymers. We will focus our attention on homologous biopolymers.

Biopolymers are synthesized inside the cells of plants and animals, usually by a condensation reaction that is catalyzed by enzymes (themselves proteins). The polymerizations are very specific, in terms of the monomers used, the distribution of the monomers along the chain (called the *sequence distribution*), and the *stereochemistry* (the three-dimensional arrangement) of the atoms in the chain. The reactions are carried out rapidly, in an aqueous environment at temperatures at or near room temperature. Unlike most synthetic polymers, some biopolymers (e.g., proteins) are *monodisperse* (all polymer chains are the same length).

Proteins

Proteins are prepared by the condensation polymerization of alpha-amino acids, as shown below:

Equation 1

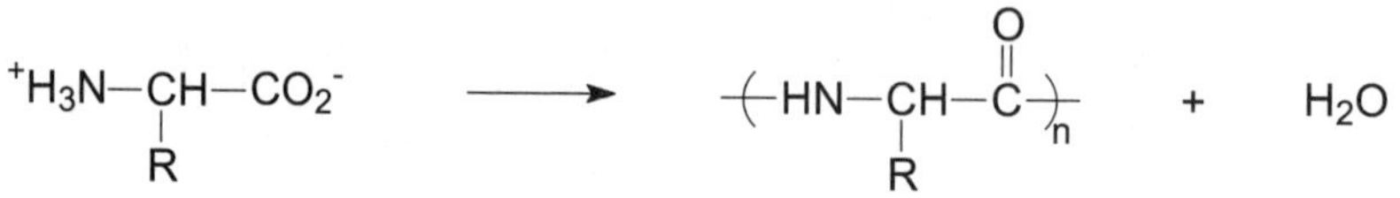

$$^+H_3N-\underset{\underset{R}{|}}{C}H-CO_2^- \longrightarrow \left(HN-\underset{\underset{R}{|}}{C}H-\overset{\overset{O}{\|}}{C}\right)_n + H_2O$$

(By convention, the α-carbon is the carbon next to the C=O.) The R in the structure stands for a hydrogen atom or an organic group. Although a few hundred α-amino acid units have been identified in nature, only 20 are used in the natural synthesis to make most proteins (see Table 3-1).

Some amino acid units other than the 20 discussed above are formed by reactions on existing protein molecules. For example, the thiol (SH) groups in two cysteine residues can be oxidized to a disulfide linkage (-S-S-) and produce a cystine unit. If the two cysteine residues are on different chains, this reaction links the two chains in what is called a chemical *crosslink*. This is shown in Equation 2. The squiggly lines indicate that the cysteine units are located along a polymer chain.

Equation 2

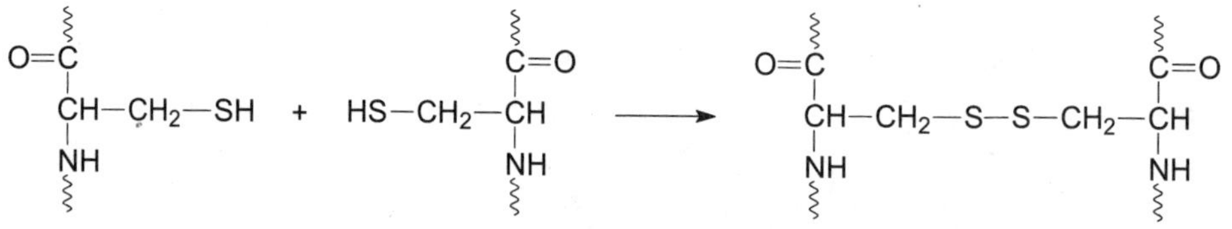

$$O=\underset{\underset{NH}{|}}{\underset{|}{C}}\;\underset{}{CH}-CH_2-SH \;+\; HS-CH_2-\underset{\underset{NH}{|}}{\underset{|}{CH}}\;\underset{}{\overset{C=O}{}} \longrightarrow O=\underset{\underset{NH}{|}}{\underset{|}{C}}\;\underset{}{CH}-CH_2-S-S-CH_2-\underset{\underset{NH}{|}}{\underset{|}{CH}}\;\underset{}{\overset{C=O}{}}$$

The bond linking amino acid units in protein molecules is an amide bond (just as in synthetic polyamides or nylons). In proteins this is called a *peptide bond* or *linkage* (see structure below). We will see that intra- and intermolecular hydrogen bonding plays a very important role in the structure and properties of all polyamides, given that the amide nitrogen is almost always bonded to a hydrogen atom. *Polypeptide* (or just *peptide*) is a term usually denoting lower molar mass polymers or oligomers, often compounds prepared synthetically. A dipeptide contains two amino acid units (one peptide linkage), an example being abbreviated Phe-Leu (see Table 3-1 for structures and names). Likewise, a tripeptide contains three amino acid residues and two linkages, as in Ala-Asp-Gly. By convention, the first amino

Table 3-1. The natural alpha-amino acids.

Structure	Name	Abbreviation	Structure	Name	Abbreviation
$H_2N-CH_2-CO_2H$	glycine	gly	$H_2N-CH-CO_2H$ $\quad\quad\mid$ $\quad CH_2CONH_2$	asparagine	asn
$H_2N-CH-CO_2H$ $\quad\quad\mid$ $\quad CH_3$	alanine	ala	$H_2N-CH-CO_2H$ $\quad\quad\mid$ $\quad CH_2CH_2CONH_2$	glutamine	gln
$H_2N-CH-CO_2H$ $\quad\quad\mid$ $\quad CH_2C_6H_5$	phenylalanine	phe	$H_2N-CH-CO_2H$ $\quad\quad\mid$ $\quad CH_2CH_2CH_2CH_2NH_2$	lysine	lys
$H_2N-CH-CO_2H$ $\quad\mid$ CH_3CHCH_3	valine	val	$H_2N-CH-CO_2H$ $\quad\quad\mid$ $\quad CH_2CH_2CH_2N=C(NH_2)_2$	arginine	arg
$H_2N-CH-CO_2H$ $\quad\quad\mid$ $\quad CH_2CH(CH_3)_2$	leucine	leu	$H_2N-CH-CO_2H$ $\quad\quad\mid$ $\quad CH_2OH$	serine	ser
$H_2N-CH-CO_2H$ $\quad\mid$ $CH_3CHCH_2CH_3$	isoleucine	iso	$H_2N-CH-CO_2H$ $\quad\mid$ CH_3CHOH	threonine	thr
$H_2N-CH-CO_2H$ $\quad\quad\mid$ $\quad CH_2CO_2H$	aspartic acid	asp	$H_2N-CH-CO_2H$ $\quad\quad\mid$ $\quad CH_2SH$	cysteine	cys
$H_2N-CH-CO_2H$ $\quad\quad\mid$ $\quad CH_2CH_2CO_2H$	glutamic acid	glu	$H_2N-CH-CO_2H$ $\quad\quad\mid$ $\quad CH_2CH_2SCH_3$	methionine	met
$H_2N-CH-CO_2H$ $\quad\quad\mid$ $\quad CH_2C_6H_4OH$	tyrosine	tyr	proline structure	proline	pro
$H_2N-CH-CO_2H$ $\quad\quad\mid$ $\quad CH_2$ 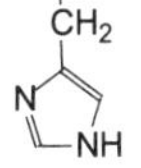 histidine ring	histidine	his	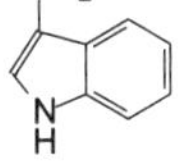 tryptophan structure	tryptophan	trp

acid unit listed has a free amino group (*N-terminal amino acid*), and the last a free carboxylic acid (*C-terminal amino acid*). Thus the structure for the tripeptide above would be written:

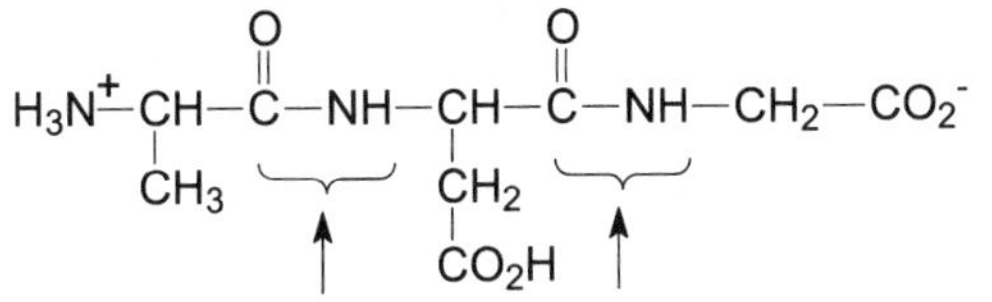

$$H_3\overset{+}{N}-\underset{\underset{CH_3}{\mid}}{CH}-\overset{\overset{O}{\parallel}}{C}-NH-\underset{\underset{\underset{CO_2H}{\mid}}{CH_2}}{CH}-\overset{\overset{O}{\parallel}}{C}-NH-CH_2-CO_2^-$$

Ala   Asp   Gly

The arrows point to the two amide or peptide links. Note that in this example, one of the amino acid residues (aspartic acid) has an extra carboxylic acid group in the side chain or R group. As a result, the molecule above is called an *acidic* tripeptide because it contains more carboxyl groups than amines.

So far, we have discussed only the *primary structure* for polypeptides and proteins, which is the identity and sequence of the amino acid groups making up the chain. The primary structure for proteins is extremely specific and is determined by the genetic code. The *secondary structure* is the shape or *conformation* the molecular chain takes in space. This can be, for example, a right-handed helix (*α-helix*) in many proteins, the result of intramolecular hydrogen bonding between different amino acid units at regular intervals along the chain. Proteins assuming this conformation tend to contain amino acid units with larger, bulkier R groups on the α-carbon and located on the outside of the helix. Commonly, the helices have approximately 3.6 amino acid units per revolution. This is shown in Figure 3-1a.

Figure 3-1. a. Alpha-helix structure for a polypeptide or protein; b. Pleated sheet structures, depicting parallel (1) and antiparallel (2) variants (Elias 1997, reprinted courtesy of Wiley-VCH.).

O   *methyl (CH$_3$) group in helices*

o   *carbon atoms in helices or CO groups in pleated sheets*

o   *hydrogen atoms in helices*

●   *nitrogen atoms in helices or NH groups in pleated sheets*

- - -   *hydrogen bonds between CO and NH groups*

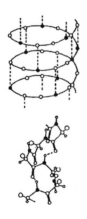

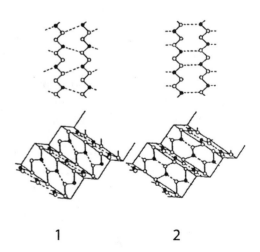

1   2

a   b

Alternatively, a protein might have *a pleated sheet* conformation, more common with sequences of amino acid units with small R groups on the α-carbon (see Figure 3-1b). Thus, the secondary structure for a given protein depends in large part upon the tendency of the R groups to attract or repel each other along the chain. In other words, the secondary structure is dependent upon the primary structure.

Tertiary structure describes more complicated conformations, perhaps involving more than one polymer chain held together by intermolecular bonds (e.g., S-S linkages; see Equation 2), ionic interactions, and/or hydrophobic interactions. (A *hydrophobic* group repels water, while a *hydrophilic* group attracts water. Therefore, hydrophobic interactions are the result of mutual attractions of hydrophobic chemical groups or molecules.) Certain hair treatments involve the deliberate manipulation of interactions between proteins. A *permanent* or *perm* involves the breaking (reduction) and reforming (oxidation) of S-S bonds in cystine units. Higher associations of molecules are called *quaternary structures*. Clearly the structures of proteins are extremely complex and are tuned for very specific applications.

Proteins can be divided into two main types, based on their overall shape. *Fibrous proteins*, as their name implies, have fiberlike structures and are used for structure or support. They are found, for example, in collagen (skin, tendon cartilage, fish scales), elastin (connective tissue), and keratins (silk, feathers, horn, and hair). They are tough and insoluble macromolecules, often having several α-helical chains wound together into ropelike bundles.

In sharp contrast, the shape of *globular* proteins is spherical or ellipsoidal. They are soluble in water and their solutions have low viscosities (see Chapter 6). They find use as transport agents, enzymes, antibodies, and some hormones. The chains in globular proteins are extensively folded and usually contain both helical and nonhelical segments.

Heating protein samples, or exposing them to solvents, can cause the breakup of quaternary, tertiary, and possibly secondary structures, causing the loss of physiological activity. This process is called *denaturation* and, in principle, is reversible. Often, however, the individual denatured molecules clump together, or *aggregate*, and then precipitate into a semisolid mass. It is usually not possible to redisperse aggregated protein. Once egg white (which contains a high percentage of the globular protein albumin) is heated slightly or exposed to acid or ethyl alcohol, it denatures irreversibly, forming an insoluble white mass.

Protein Synthesis

The *biosynthesis* of proteins is the subject of biochemistry and will not be covered here. In the laboratory, if we were to heat a mixture of α-amino acids together in an attempt to form a polypeptide, the result would be only a random copolymer. To synthesize a peptide or protein, we would have to introduce each amino acid residue, one at a time, in the proper order. Given that the molar mass of many proteins exceeds 100,000, such a manual synthesis would be extremely tedious. In the early 1960s, Bruce Merrifield of Rockefeller University introduced an ingenious method for doing this. Although the process is chemically a bit more complicated than what is described here, the general outline follows. Merrifield prepared small beads of polymers containing functional groups that react with the amino groups of amino acids (*bead polymerization* is covered in Chapter 5). Adding these beads to a solution of the first amino acid of a peptide anchors that amino acid to the beads. The beads are isolated by filtration, washed to remove excess amino acid, and then added to a solution of the next amino acid in the sequence. After the last amino acid has been added, the bond holding the peptide to the bead is broken, releasing the final product. The process is shown schematically for the synthesis of the peptide Ala-Asp-Gly:

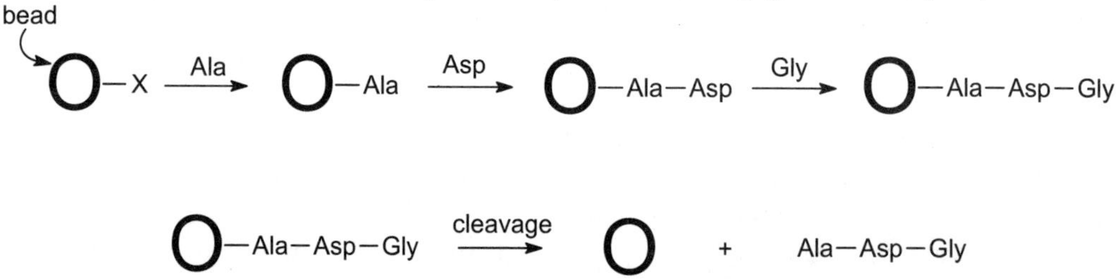

In this way, it is possible to synthesize peptides with the desired sequence of amino acids. Although the process has been completely automated, this approach is only used to prepare peptides of relatively low molar mass. Merrifield's approach, called *solid-phase synthesis*, has been expanded as a more general tool for synthesizing a wide range of organic compounds.

Some Specific Types of Proteins

Enzymes are generally globular proteins whose primary function is the catalysis of biochemical reactions. Their names, which end in "ase," indicate their specific function. For example, glucose polymerase is the catalyst for polymerizing glucose. Their overall shape includes a substrate binding site, in which the reacting molecule or portion of a molecule can fit, and a catalytically active site at which the reaction occurs. Although catalysis in nature is their most important function, some enzymes are prepared by microorganisms in reactors for commercial uses, such as for medical or biochemical applications. Some are used by the food industry—for example, in cheese making.

Wool, which is the hair of sheep and goats, contains more than 200 different compounds, of which approximately 80% are keratins (fibrous, helical proteins). Two or three helices are chemically interconnected or crosslinked by, for example, S-S bonds to form a protofibril. Eleven protofibrils make up a microfibril (approximately 5 nm in diameter), and several microfibrils form a macrofibril. A number of macrofibrils are incorporated with other structures into a wool fiber, which is on the order of 10 to 200 μm in diameter. The complexity of wool fiber is illustrated in Figure 3-2.

Silk is produced by a number of organisms, including spiders, silkworms, scorpions, mites, and flies (Kaplan et al. 1997). The silk from the cocoon of the mulberry silk spinner caterpillar (*Bombyx mori*) has been the most important source of natural silk for textiles for more than 5000 years. The cocoons consist of about 80% silk fiber and 20% glue. After immersion in hot water to soften the glue, the cocoons are treated with mechanical brushes that unwind the fibers. The process of raising the caterpillars, harvesting the cocoons, and obtaining the silk fibers requires much hand labor, meaning that items made of silk are quite expensive relative to those made of other natural and manmade fibers. Because of its unique luster (the way it reflects light) and its feel, silk remains a very desirable material for clothing. As we will see in the next chapter, women's stockings made from silk were both coveted and loathed prior to the introduction of "nylons," the synthetic replacement with superior properties. They were coveted because there was no substitute and loathed because they sagged and developed runs very easily. They were an expensive and necessary dress item in most women's wardrobes.

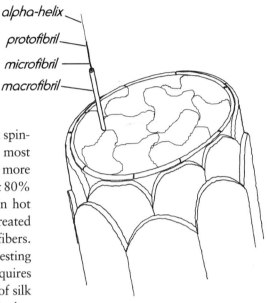

Figure 3-2. Schematic drawing of a wool fiber, showing alpha-helix, protofibril, microfibril, and macrofibril (Drawing courtesy of Stephen Teegarden).

alpha-helix

protofibril

microfibril

macrofibril

Silk from the Spider?

We've all been intrigued at one time or another by a spider's web. You might be surprised to learn that some spiders actually spin up to six or seven different kinds of silk, each with its own set of mechanical properties and purposes. One is for wrapping and protecting eggs, another is used to tie up prey, and three or four different ones can be used in the construction of webs. Spiders use dragline silk as a safety line and to frame their webs. Dragline silk is some three times stronger than polyaramid fibers, used to make bulletproof vests, and five times stronger than steel, by mass (Lazaris et al. 2002). Analysis of these silk fibers reveals alternating alanine-rich blocks and glycine-rich blocks (see Table 3-1). The former are crystalline (see Chapter 7), which provide great strength, while the latter provide elasticity, allowing the silk fiber to stretch.

How Do They Make It?

Spiders have cells that synthesize and secrete two soluble proteins into an acidic water-based solution contained inside the silk gland. The concentration of the polymers in the gland is 30% to 50% by mass. When the proteins are ejected from the gland, they self-assemble into long fibers, approximately 20 mm in diameter for dragline silk (Lazaris et al. 2002; Service 2002). Note that although the initial protein molecules are soluble in water at low pH, once spun into fibers they are insoluble. Not only is this dual property essential for the preparation and ultimate function of the silk, it points out rather dramatically the significance of the secondary, tertiary, and quaternary structures of proteins. Clearly the silk in a spider's web cannot redissolve, because it must survive humid conditions and rain storms. Interestingly, however, silk does break down in the environment over time (is *biodegradable*; see Chapter 9).

How Can We Get It?

Given the strength of dragline silk fiber, there has been considerable interest for many years in obtaining it in large quantities for industrial and medical applications. For example, it would be very useful as, say, strong, biodegradable suture material for eye surgery. Many efforts, for more than a century, to raise spiders and collect their silk have all failed. Spiders in captivity tend to eat each other. To produce silk fibers synthetically, two basic hurdles must be overcome. First, protein molecules with the correct primary structure and with high molar mass (on the order of 150,000) must be synthesized. Next these molecules (usually pairs of molecules in the spider) must be extruded into fibers. As we will see throughout this text, *synthesis* and *processing* are the two most basic functions of polymer science. For several years, various scientists have attempted to use genetic engineering to transfer silk-making genes into other organisms. Recent efforts using the lactation system in cows and goats have met with modest success (Lazaris et al. 2002; Robbins 2002). The gland that produces milk in mammals is similar to that which the spider uses to produce silk. By using *biomimicry*, the adaptation of natural phenomena to produce synthetic materials, scientists have been able to produce silk fibers with mechanical properties that are beginning to approach those of natural silk.

Polynucleotides
Chemical Structure

Polynucleotides (also called *nucleic acids*) are polyesters of phosphoric acid and sugar molecules in which each sugar unit has a base bonded to it:

—phosphate—sugar—phosphate—sugar—phosphate—sugar—phosphate—sugar—
　　　　　　　|　　　　　　　　　|　　　　　　　　|　　　　　　　　|
　　　　　　base　　　　　　base　　　　　　base　　　　　base

The two major types of sugar units are D-ribose and D-2-deoxyribose, the latter having one fewer OH group than does ribose:

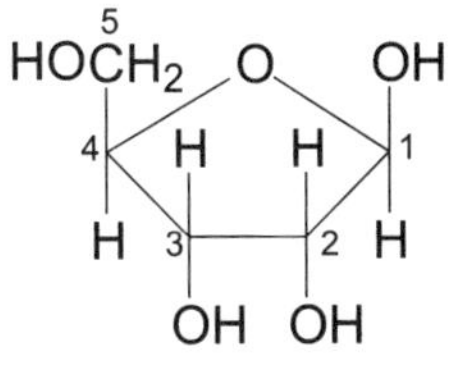

HOCH₂ ... D-ribose   HOCH₂ ... D-2-deoxyribose

D-ribose

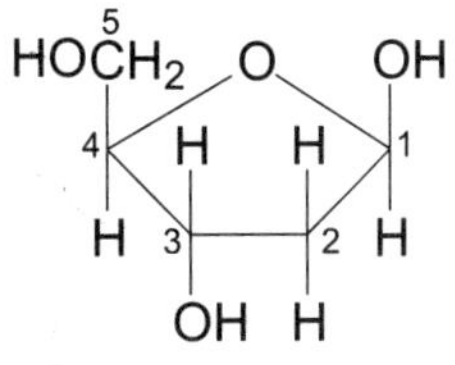

D-2-deoxyribose

Each molecule contains a five-membered ring consisting of one oxygen and four carbon atoms. To make the structure less cluttered, the carbon atoms in the ring (numbers 1 through 4) are not shown but are understood to be at each angle in the line. The bases are connected to the sugar molecules at carbon 1, replacing the OH group.

In the polymers, the sugars are bonded to phosphate through the oxygen atoms on carbons 3 and 5. Polynucleotides containing D-ribose are called *ribonucleic acids* (RNA), while those containing D-2-deoxyribose are called *deoxyribonucleic acids* (DNA). Putting this all together, we can draw structures for RNA and DNA chains, as depicted in Figure 3-3.

The structures of the most important bases are drawn on p. 36. They are all cyclic, nitrogen-containing compounds. The hydrogen atoms with the circles around them indicate the specific nitrogen in each base that is bonded to a sugar molecule.

Figure 3-3. Segments of RNA and DNA polymer chains.

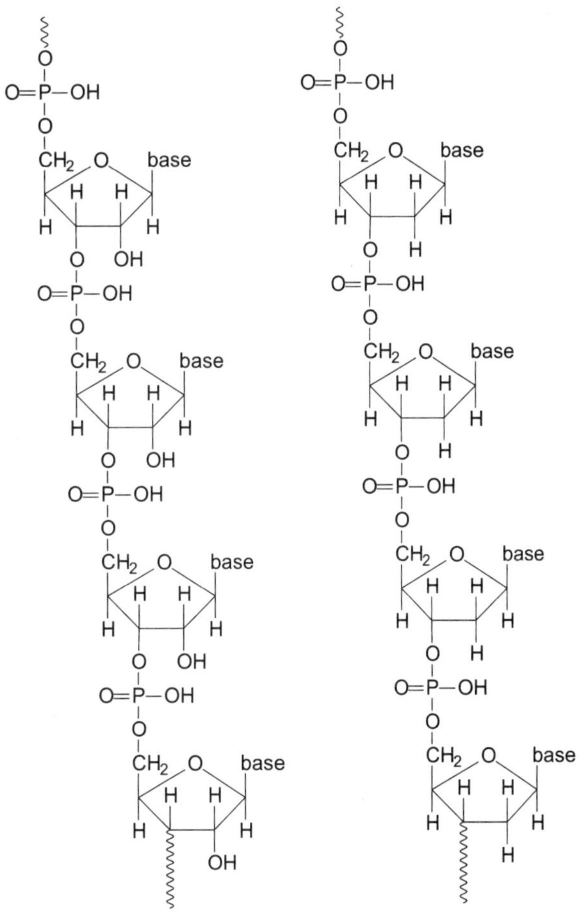

portion of RNA chain   portion of DNA chain

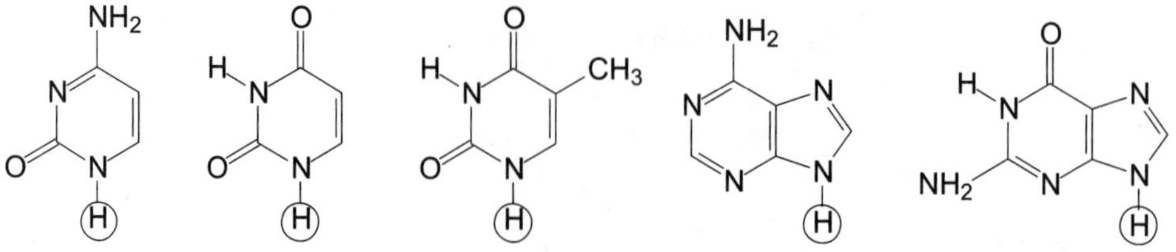

| cytosine | uracil | thymine | adenine | guanine |

Although their primary structures seem quite similar, RNA and DNA are really very different polymers. The molar mass of DNA is considerably higher, often being in the hundreds of millions. The molar mass of DNA in the bacterium *Escherichia coli* is some 2.5×10^9 (Stevens 1999). If the polymer were stretched out, it would be several millimeters long. To fit into a cell nucleus approximately 5 micrometers in diameter, the DNA molecule must be extensively folded.

The four most important bases occurring in DNA are cytosine, thymine, adenine, and guanine. The composition of these bases in DNA is very regular. For complementary strands, the number of cytosine (C) units equals the number of guanine (G) units, while the number of thymines (T) equals the number of adenines (A). The regularity of the composition of the polymer gives rise to regularity in its secondary and tertiary structures. It was this base regularity, combined with the X-ray images obtained by Rosalind Franklin, that led James D. Watson and Francis H. C. Crick to propose that the structure of DNA in higher life forms is a double helix (Watson 1968). That is, two separate, complementary polymer chains of DNA, running in opposite directions, twist around each other in a right-handed helix. The structure is stabilized by intermolecular hydrogen bonding between the matched base units (C paired with G, and T paired with A). In addition to maintaining the regularity of the double DNA strands, the base pairs play another critical role—it is the *sequence* of these pairs that determines whether a particular organism is, for example, a bacterium or a human being.

Biochemical Function

The DNA that is twisted in a double helix and folded into a cell nucleus is called a *chromosome*. DNA is the chemical species that contains the genetic information necessary to make protein to duplicate a cell. It is the blueprint for all of the protein in the entire organism. Each chromosome contains hundreds to thousands of *genes*, short sequences of DNA that function as code for particular amino acids. Human DNA has approximately 30,000 genes. A *genome* is the sum of all of the DNA in an organism. The human genome has some 3×10^9 pairs of bases. In each gene, a set of three bases signifies one amino acid. For example, the sequence CGG is the code for the amino acid arginine. During protein synthesis, when this sequence is encountered, an arginine molecule will be incorporated into the growing protein.

During cell division (replication), part of the double helix unwinds, temporarily separating the base pairs. The individual strands then act as templates for producing new, complementary strands. After replication, each double helix contains one old and one new strand, thereby ensuring that all genes are intact. DNA is responsible also for the synthesis of RNA.

Besides having a much lower molar mass than DNA, RNA generally forms only single-strand helices. RNA is often found associated with proteins inside cells. The most prevalent bases in RNA are the same as those in DNA, except that uracil is present instead of thymine. Three common types of RNA are ribosomal (rRNA), transfer (tRNA), and messenger RNA (mRNA). They are all involved in protein synthesis, controlling the sequence of amino acids that make up the primary structure. Thus the base sequence in RNA is related to the amino acid sequence in the protein that is made from it.

Genetic engineering involves the alteration of DNA in a cell with new polynucleotide segments, resulting in *recombinant* DNA. One can use the cells of a host organism containing recombinant DNA to produce protein useful in medicine such as insulin.

Polysaccharides

Polysaccharides are the most prevalent of the biopolymers, with cellulose alone making up approximately one-third of the total solid matter in the entire plant kingdom. It is the most abundant organic material on earth (Goldstein 1977). See Table 1-1 for the main classes of these important polymers. We will see in some instances that the monomer units making up different polysaccharides are essentially identical, even though the properties of the polysaccharides are vastly different. The explanations for this are very interesting and remind us yet again of the importance of *stereochemistry* in nature.

Structure

The building blocks for polysaccharides are *monosaccharides* (sugars) such as glucose or fructose ($C_6H_{12}O_6$). Two cyclic structures of D-glucose are reproduced below (a third, open-chain structure is not shown).

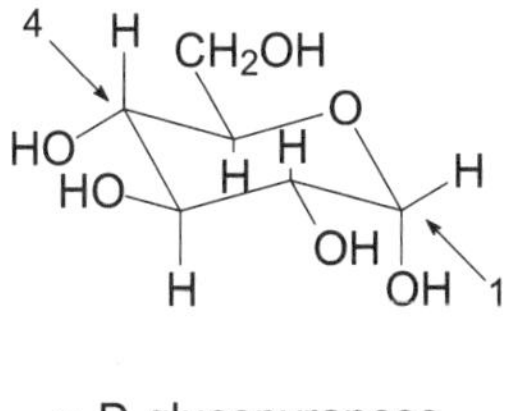

α-D-glucopyranose 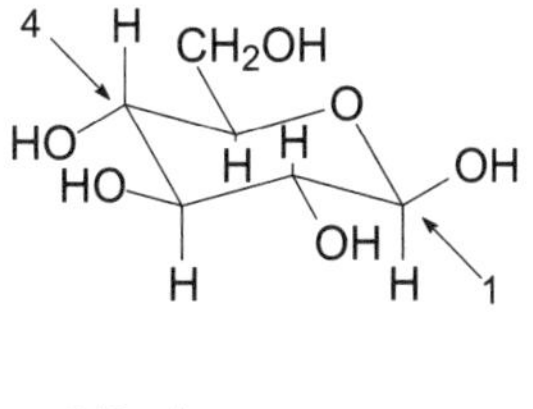   β-D-glucopyranose

The arrows indicate carbon atoms 1 and 4. These two structures are identical except for the positions of the hydrogen and the OH groups at carbon number 1. When the OH on carbon-1 is down, the stereochemistry is called *alpha* ("α"), and when it is up, *beta* ("β").

Cellobiose and maltose are both *disaccharides* made up of two glucose units. The bond linking the two glucose units is called a *glycosidic linkage*, indicated by the arrows in the following structures:

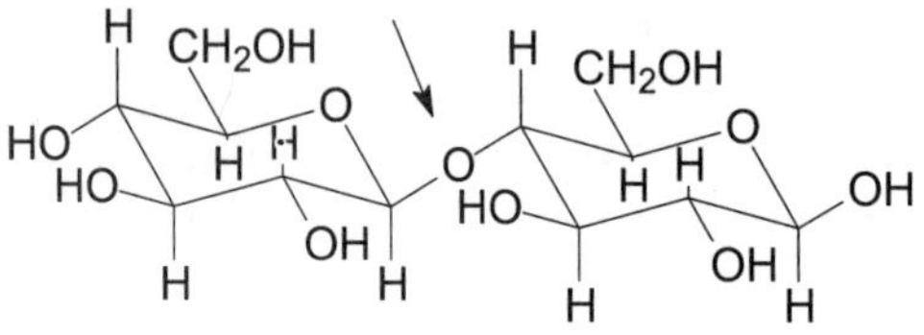

cellobiose

(β-glycosidic linkage)

maltose

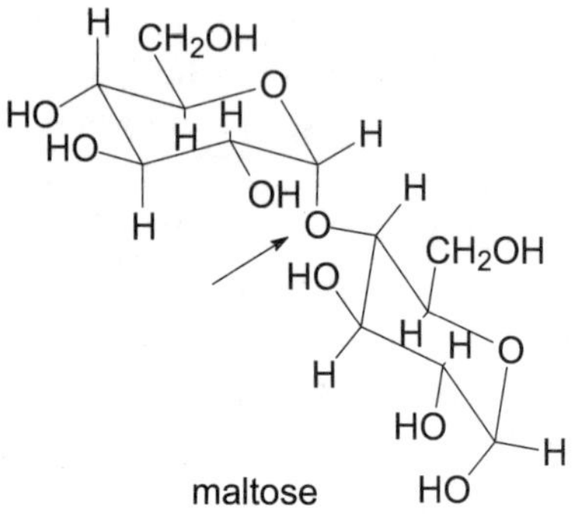

(α-glycosidic linkage)

Note that in both cellobiose and maltose, the new bond, the glycosidic bond, is between carbon-1 of one molecule and carbon-4 of the other. In cellobiose, however, it is specifically a β bond, while in maltose the bond is α. *Cellulose*, the *polysaccharide*, is made up of some 3500 glucose units all bonded together through carbon atoms 1 and 4 and all with β glycosidic linkages, just like cellobiose. The result is a linear polymer chain that is very regular and "straight." Note that we refer to the chain of cellulose as being linear, even though it is made up of 6-membered rings. Here, linear refers to a "straight" rather than to a branched chain. The long chains hydrogen bond with each other, forming a crystalline matrix that is insoluble in water.

In sharp contrast, *starch* and *glycogen* are polysaccharides that either are water-soluble or can be dispersed in water. They have none of the structural rigidity of cellulose. How interesting that both starch and glycogen are also homopolymers of D-glucose. And like cellulose, they have glycosidic bonds between carbons 1 and 4. However, the structure at carbon 1 is α, not β as in cellulose. So this "slight" difference in stereochemistry leads to a profound difference in properties.

Starch exists in two major forms, *amylose* and *amylopectin*. Although amylose is a linear polymer like cellulose, the α stereochemistry enables the chains to twist and to assume a helical shape. Small molecules such as I_2 fit inside the helix and form a complex. This is the basis for the deep blue color in the starch-iodine test. This "non-straight" or "bent" conformation is suggested in the structure of the disaccharide maltose, shown above. Amylose makes up approximately 25% of many starches and has molar mass in the range of 100,000 to 750,000. In contrast to the linear amylose, amylopectin is highly branched, making it soluble in water. The branches are on carbon number 6, and molar mass can exceed 1 million. Glycogen is similar to amylopectin, except that glycogen is considerably more highly branched.

The Color of Money

Did you ever pay for an item with an American $20 bill and watch the clerk make a mark on it with a felt-tipped pen? The clerk is determining whether the bill is actual currency or counterfeit. The special pen that the clerk uses contains an iodine solution. I_2 forms complexes inside the helices of the polysaccharide amylose that is a component of starch.

Paper is made from cellulose, either from cotton fibers or from wood that has been converted to pulp. Because ordinary paper contains a certain concentration of starch, contacting it with a counterfeit-detecting pen will leave a blue-black mark. Since 1959, U.S. paper currency has contained virtually no starch. Therefore contacting it with a special pen leaves only a golden-brown mark.

Counterfeiters often use inferior grades of paper, including paper that contains starch. Thus this simple chemical test can be an accurate technique for determining forgeries. Modern U.S. paper currency (other than $1 bills) contains other features that make it difficult to counterfeit. The ink is magnetic. You can probably detect this by holding a bill at the top and allowing the bottom to approach a very strong magnet such as a neodymium one.

Also, newer bills contain a plastic strip that is visible when the bill is held up to a light. The strips fluoresce in ultraviolet light: $5—blue, $10—orange, $20—yellow-green, $50—yellow-orange, and $100—pink.

Function and Properties

So what effect do these differences in stereochemistry and extent of branching have on the uses of these polysaccharides? Because of the extended nature of its chains, cellulose is the most common *structural* component of plants. Cellulose makes up the main component of the cell wall of plants. Wood is approximately 50% cellulose, while fiber-producing plants such as flax, jute, and hemp are 65% to 80%. The seed hairs of cotton are virtually pure cellulose. The long fibers from cottonseed (up to 5 cm long and 9 to 25 μm in diameter [Franz 1986]) are spun into thread that is then woven into fabric for clothing. Short fibers, called *linters*, are used as the raw materials for chemical derivatives such as cellulose acetate (see Chapter 4).

In stark contrast, starch is primarily used for energy storage in plants and is therefore a vital source of dietary carbohydrates for animals, including humans. It serves as a link in the photosynthesis chain, allowing organisms that do not photosynthesize, such as mammals, to utilize energy from the sun. Tuberous plants such as potatoes are plentiful sources, as are grains such as wheat, corn, and rice. Given its structure, starch is much more readily hydrolyzed than cellulose and is a critical biopolymer in most plants. In animals, glycogen is the primary energy storage polysaccharide. Primates are unable to metabolize cellulose, because they lack the enzymes necessary to hydrolyze it. Cows and other ruminants do utilize cellulose as food because of the presence of symbiotic bacteria in their gut.

An Ocean of Polysaccharides

Plants and animals present us with a rich variety of polysaccharides in addition to those discussed above. Seaweed is a source of several, including the alginates, agars, and carageenans. Alginic acid is the major cell wall component in brown algae, a readily available raw material (Franz 1986). While sodium alginate is soluble in water, calcium alginate is not and forms a gel. This property makes sodium alginate useful as a food additive. For example, if large drops of a mixture of pureed cherries and sodium alginate are added to a Ca^{2+} solution, imitation "cherries" are formed that are commonly used in baked goods. Alginates and carageenans are also used as thickeners in a number of food products such as ice cream.

The exoskeletons of shellfish such as crabs and lobsters as well as many insects contain a high concentration of *chitin*, which is very similar to cellulose (b 1,4-glycoside linkages), except that instead of an OH group at carbon-2, chitin has a substituted amine group (an amide):

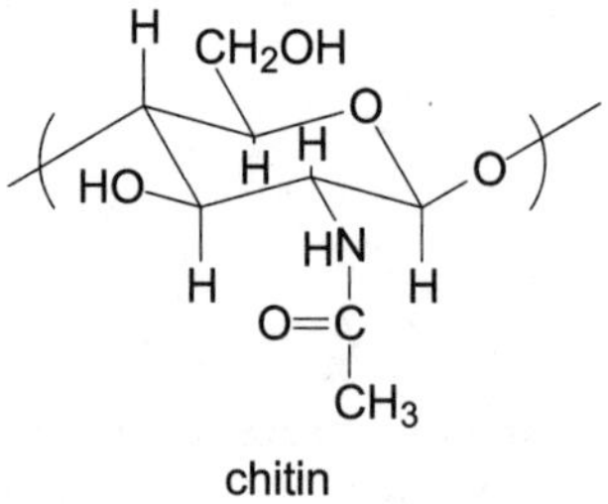

chitin

Like cellulose, the primary biological function of chitin is for structural support. And like cellulose, it is quite insoluble. Strong acids hydrolyze the amide to give the protonated amine, which is soluble at low pH. Given its abundance in nature, there has always been interest in finding commercial application for chitin or its derivatives. So far these efforts have achieved only limited success.

Other polysaccharides have found widespread application. Gums, which are complex, highly branched polysaccharides produced by plants, form very viscous solutions in water at low concentrations. Examples such as *gum tragacanth* have been used as thickening agents in foods in place of starch and to alter the texture of ice cream.

Natural Rubber

A number of plants and some trees contain a white, milky liquid that is released when the stem or bark is cut. The liquid is called a *latex* from the Latin meaning "liquid." Common sources include dandelions, milkweed, goldenrod, and potted rubber plants. Rubber trees, from which substantial quantities of latex can be harvested, grow in some tropical areas of the world. A major constituent of this latex is a homopolymer of isoprene (2-methyl-1,3-butadiene), called *polyisoprene*. Polyisoprene, as well as a number of other elastomers, has a carbon-carbon double bond in every repeat unit. The properties of polyisoprene are the result of the presence of these double bonds. Just as stereochemistry plays a critical role in both proteins and polysaccharides, we will see its importance here.

The stereochemistry in a polymer such as polyisoprene arises because of the rigid nature of the carbon-carbon double bond. Each double bond in the polymer chain can exist in one of two possible *stereoisomers* as show below. Stereoisomers are compounds that differ from each other only in the way their atoms are oriented in space. If two similar groups are on the same side of the double bond (here the similar groups are the carbon atoms in the backbone of the polymer), the stereoisomer is called the *cis* isomer. If the two similar groups are on opposite sides of the double bond, the stereoisomer is called *trans*.

$$\left(\begin{array}{c} CH_2 \\ H_3C \end{array} C=C \begin{array}{c} CH_2 \\ H \end{array} \right)_n \qquad \left(\begin{array}{c} CH_2 \\ H_3C \end{array} C=C \begin{array}{c} H \\ CH_2 \end{array} \right)_n$$

cis-polyisoprene   trans-polyisoprene

(natural rubber)   (gutta percha)

Polymer with the cis stereochemistry, which is obtained primarily from the *Hevea brasiliensis* tree, is called *natural rubber*. *Hevea* originated in Brazil, from where several thousand seedlings were taken in the late 19th century and eventually transplanted in what was then Ceylon (now Sri Lanka). Today at least three-quarters of the world's natural rubber comes from plantations in Malaysia, Indonesia, Thailand, and Sri Lanka. The name "rubber" is attributed to Joseph Priestley, the British chemist well known for discovering oxygen. He called the material "India rubber" because it came from the East Indies and it was "excellently adapted to the purpose of wiping from paper the unwanted marks of a black lead pencil" (Fenichell 1996). Americans call such an item an "eraser," while the British still refer to it as a "rubber."

Gutta percha is obtained from the *Palaquium oblongifolium* tree in Malaysia, Indonesia, and South America. Unlike the very soft natural rubber, gutta percha is a hard substance at room temperature that softens on heating. Its trans geometry results in a more linear, extended structure, and as a result the chains pack together and crystallize more readily (see Chapter 7). Although once widely used in golf balls and chewing gum, gutta percha has now been largely replaced by synthetic polymers.

A third form of natural polyisoprene is *balata*. It is similar to gutta percha in that its stereochemistry is also primarily trans. It is obtained from the *Mimusops globosa* tree (the bully tree) that grows in Venezuela, Barbados, and Guyana. Balata contains only about 50% polyisoprene, the remainder consisting primarily of an ill-defined mixture of organic resins. It is still used as the cover material in some high-end golf balls.

At least one natural polyisoprene consists of both cis and trans isomers. *Chicle*, found in Central America, is approximately 75% trans, the remainder being cis. It was used as the base for chewing gum for many years, until it was replaced in the midtwentieth century by the synthetic polymer poly(vinyl acetate).

Proteins, nucleic acids, and polysaccharides all serve critical biological functions. They are synthesized for specific purposes and are essential to the lives of the organisms in which they occur. The role of the polyisoprenes, on the other hand, is less obvious and less well understood. Because their backbones contain a large number of double bonds, they decompose readily when exposed to sunlight or to oxygen in the air. Therefore, rubber did not become a widely useful material until the invention of vulcanization in the 1850s by Charles Goodyear and Thomas Hancock (see Chapter 4).

Summing Up

Nature provides us with a fascinating array of biopolymers, fulfilling a range of functions critical to the support of life. Some provide structure to cells, to plants, or to animals. Some are used for warmth or protection from the elements. Others are part of a food chain, control biochemical reactions, or store and transfer genetic information. The function of natural macromolecules is very much dependent upon their shape, and this is determined to a large degree by stereochemistry. Sometimes subtle changes in stereochemistry cause huge differences in the properties of natural polymer. Both proteins and polysaccharides rely extensively on hydrogen bonding, either within the same molecule or between molecules, to determine the shape or conformation of their chains. The secondary and tertiary structure of proteins depends heavily on the hydrophobic and hydrophilic interactions within a single chain, as well as on the formation of disulfide linkages from the oxidation of two cysteine units.

The laboratory synthesis of many natural polymers is extremely challenging. As we will see in Chapter 5, methods for controlling stereochemistry during polymerization are still somewhat limited.

References Cited

Elias, H-G. 1997. *An introduction to polymer science*. 154. New York, NY: Wiley-VCH.

Fenichell, S. 1996. *Plastic: The making of a synthetic century*. New York, NY: HarperBusiness.

Franz, G. 1986. Polysaccharides in pharmacy. In *Pharmacy, thermomechanics, elastomers, telechelics*, ed. K. Dusek, 11. New York, NY: Springer-Verlag.

Goldstein, I. S. 1977. The place of cellulose under energy scarcity. In *Cellulose chemistry and technology*, ed. J. C. Arthur, Jr., 382–87. Washington, DC: American Chemical Society.

Kaplan, D. L., et al. 1997. Silk. In *Kirk-Othmer encyclopedia of chemical technology*, ed., M. Howe-Grant, 4th ed., vol. 22, 155–63. New York, NY: John Wiley and Sons.

Lazaris, A., et al. 2002. Spider silk fibers spun from soluble recombinant silk produced in mammalian cells. *Science* 295: 472–76.

Robbins, J. 2002. Second nature. *Smithsonian* 33 (4) (July): 78–84.

Service, R. F. 2002. Mammalian cells spin a spidery new yarn. *Science* 295: 419–20.

Stevens, M. P. 1999. *Polymer chemistry: An introduction*. 3rd ed., 505. New York, NY: Oxford University Press.

Ward, W. H., and R. A. O'Connell. 1986. Wool. In *Encyclopedia of materials science and engineering*, ed., M. B. Bever, vol. 7, 5461. Cambridge, MA: The MIT Press.

Watson, J. D. 1968. *The double helix: A personal account of the discovery of the structure of DNA*. New York, NY: Atheneum.

Other Reading

Allcock, H. R., and F. W. Lampe. 1990. *Contemporary polymer chemistry*. 2nd ed., ch. 8, 162–95. Englewood Cliffs, NJ: Prentice-Hall.

Berg, J. M., J. L. Tymoczko, and L. Stryer. 2002. *Biochemistry*. 5th ed. New York, NY: W. H. Freeman.

Coultate, T. P. 1989. *Food: The chemistry of its components*. 2nd ed. London: The Royal Society of Chemistry.

Elias, H-G. 1987. *Mega molecules*. New York, NY: Springer-Verlag.

Horton, H. R., L. A. Moran, R. S. Ochs, J. D. Rawn, and K. G. Scrimgeour. 2002. *Principles of biochemistry*. 3rd ed. Upper Saddle River, NJ: Prentice-Hall, Inc.

Kamenetskii, F. 1993. *Unravelling DNA*. New York, NY: Wiley-VCH.

Widmaier, E. P. 2002. *The stuff of life: Profiles of the molecules that make us tick*. ch. 1. New York, NY: Henry Holt and Company.

A History of Polymers: The Road to the Plastic Age and Beyond

> *Research is to see what everyone else has seen and think what no one else has thought.*
> Albert Szent-Gyorgyi, Nobel Laureate

45

To try to appreciate the development of polymers to the fullest, it might be best to think of the discipline of physical science as a tree. An oak will do. A great oak starts out as a small sapling, not very impressive and with leaves a little too big for its slender twigs. Slowly, ever so slowly, it grows and develops new branches. At the start of the twentieth century, the tree of physical science was a respectable specimen, with stature, deep roots, and lots of branches. Over the next 100 years, it would continue to grow and expand, although no one could envision just how. One hundred years ago the large branches that would become known as polymer science were mere offshoots of more established branches. Polymers were thought to be unique collections of small molecules, not large *macro*molecules. No one could envision an entirely separate field devoted to synthetic polymers—there was thought to be no such thing at the time. Also lacking were branches that represent the large body of sophisticated experimental techniques we take for granted today.

So we need in this chapter to try to understand this transition, the struggles the field of polymer science underwent, and the arguments, often heated, that the word "macromolecule" generated. And we need to gain an appreciation for the courage of a few visionaries, whose intelligence and perseverance got us where we are today. Because any new field that throws conventional wisdom into upheaval goes through a period of denial, rebuke, and uncertainty. That's just human nature. We need to try to understand both the development of the synthetic polymers and the personalities of some of the people who were instrumental in their development.

We now well appreciate, of course, that polymers are virtually everywhere. Some of them occur naturally, and we continue to better understand their compositions, structures, and properties. Many of these materials have been used since the dawn of human existence, for food, obviously. Cellulose alone has been essential for clothing, fire, shelter, tools, weapons, writing, and art. Leather is probably the result of the first synthetic polymer reaction, essentially the crosslinking of protein (elastin). How we progressed over time to the "Polymer Age" is a fascinating series of stories, some of which are well worth recounting here.

Science is the endeavor to understand and to explain. Before the emergence of a scientific discipline, especially in the physical sciences, a significant period of technological advance has already occurred, during which tools, materials, and processes have been developed, refined, and perhaps commercialized. Such activity begins to draw the attention of scientists, who may or may not have any interest in the technological or commercial aspect of an area. Their primary contribution is the application of the scientific method, asking the right questions, performing the careful and systematic experimentation that leads to understanding, hypotheses, and theories. The individual motivated to produce and sell a product or material usually lacks the time, and sometimes the interest, in trying to understand "why." Given the involvement of scientists, we have the genesis of a truly scientific endeavor, which is sometimes catalyzed by a bold step in an unexpected direction.

Before Synthetic Polymers

So when did *polymer science* begin? The answer is not easy to determine, so it will help to attempt to understand why. Let's try to imagine our world without synthetic polymers. Our clothing would consist entirely of cotton, wool, linen, and, if we could afford it, silk. Our shoes would be made of leather. Raincoats would consist of cotton coated with natural rubber. Automobiles would be considerably heavier, because they would be made mostly of steel, glass, and natural rubber. Their interiors would be covered in leather. They would contain no air bags. Tires would not last very long, would puncture easily, and would be very expensive.

Sporting equipment would be much more primitive. Skis would be wooden and ski boots, leather. Ski clothing, too, would be heavy and bulky, because it would be made of wool—no polypropylene to wick perspiration away from the body, no fleece, or breathable, waterproof expanded polytetrafluoroethylene (PTFE) such as GoreTex. Baseball, football, and soccer would be played outdoors in stadiums with real grass, a situation that some would prefer. The balls would get quite heavy in the rain, because they would be covered in leather. Football helmets, too, would be leather and would not offer much protection. A good golfer would drive a golf ball about 40 to 50 yards fewer. Cameras would be very heavy and awkward. For each picture, the photographer would have to insert a glass plate on which the photographic emulsion is coated, snap the picture, then remove the plate and insert another. The camera body would be made out of wood, metal, or a composite of a natural resin such as shellac and cardboard. So you could enjoy still pictures, but there would be no motion pictures, no movie theaters.

In the grocery store, meat, fresh vegetables and fruits, and bread would be considerably more expensive, would be produced nearby, and would be less readily available out of season. Much more food would spoil because it would be packaged poorly or not at all. Soft drinks, juices, and bottled water would be packaged only in glass, steel, or aluminum. Bags of potato chips, popcorn, and snacks would be much more expensive, because they would be sealed in aluminum foil. At the checkout counter, no one would ask, "Paper or plastic?" And we would have to be prepared to pay considerably more. All of the fast-food restaurants, in addition to McDonald's, would serve everything in paper containers, including the coffee. We would pay for everything with cash or check—no "plastic"! No ATMs, either, so we would need to get to the bank before it closed.

The floors of our homes would be covered with expensive wool carpets, cotton throw rugs, linoleum, or hardwood. Paint would all be oil-based and would not last very long, especially outdoors, and paintbrushes would all be natural bristle. Forget vinyl siding, vinyl shutters, vinyl windows, polyurethane insulation, and silicone caulk. In the bathroom, you would be careful with that glass bottle of shampoo—if you dropped it, it would break when it hit the porcelain tub. At least you wouldn't spend time cleaning contact lenses—you wouldn't have any. Your glasses would be heavy—glass lenses—and would have only metal frames. You would brush your teeth with a wooden-handled toothbrush containing natural bristles.

There would be no cell phones and no instant messaging or e-mail, because there would be no home computers. No CD players, either, because there would be no CDs. In fact there would be nothing with integrated circuits. Television sets would be very large and heavy and would not produce a picture until the vacuum tubes had warmed up.

Compact Discs—Making Enough Lexan for Ludwig

If you are young, it probably seems as if audio compact discs (CDs) have been around forever. It might surprise you to learn they have been with us for only about two decades. It also might surprise you to learn that the first pop music CD available was ABBA's album *The Visitor*, released in 1982. That is a piece of trivia most of your friends probably don't know.

The CD was developed as a technology for reading digital audio signals optically, thereby eliminating the surface noise and imperfections inherent with "vinyl" recordings and magnetic tape. Commercializing the technology had to await the development of inexpensive and efficient lasers that were required for reading the data. Phillips, the Dutch electronics corporation, collaborated with Sony to set standards and codevelop the hardware.

To be successful, compact discs needed to be strong, lightweight, and extremely pure optically. Only the plastic known as *polycarbonate* met all of the requirements, which included that of maintaining its shape at elevated temperature and humidity. General Electric Corporation is a major manufacturer of polycarbonate, which they call Lexan. If you remove a CD from the glove compartment of a car that has been sitting in the sun on a very hot day, you want it to play flawlessly right away. The laser must read the microscopic signals whether the disc is hot or cold. Any slight haziness or impurity in the polycarbonate layer can deflect or interfere with the laser beam, leading to errors and therefore distortion of the audio signal. We will discuss the properties of the polymers involved in Chapter 7.

In the years since the introduction of the first CD, the lasers and other hardware have changed significantly. We now have read-only memory discs (CD-ROMs) for the optical storage of many different kinds of data, CDs for our computers that we can

Cross section of audio CD, label side on top (not to scale)

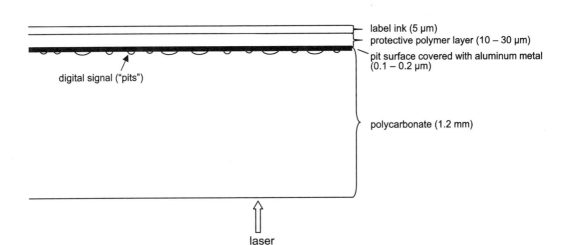

label ink (5 µm)
protective polymer layer (10 – 30 µm)
pit surface covered with aluminum metal (0.1 – 0.2 µm)

digital signal ("pits")

polycarbonate (1.2 mm)

laser

write to (or *burn*) (CD-Rs), and DVDs (digital versatile discs). Throughout all of these developments, the only base material still used is the polymer polycarbonate.

Here's another piece of trivia: Why are compact discs 12 cm in diameter? Apparently, Phillips had originally planned to make the discs 11.5 cm wide, providing a maximum playing time of 60 minutes. Today it is not known for certain who suggested that they be made slightly larger. The reason, however, was to fit Beethoven's (his first name was Ludwig) *Ninth Symphony*, approximately 74 minutes long, on one disc. And so this became the standard.

Sound primitive? Certainly by today's standards, life would be very different and much less convenient. We would be much more reliant on renewable resources such as cotton, wool, and wood. Some would argue that decreasing our reliance on nonrenewable resources such as petroleum or coal would be a step in the right direction. However, without synthetic polymers our society would rely much more on agriculture to provide not only our food but also all of our fiber for clothing, linens and bedding, upholstery, and carpets. We would have to grow more food to make up for tremendous increases in spoilage. More significantly, we would need huge areas of land to provide enough cotton for clothing and enough grazing land for the millions of sheep that would be needed for their wool. It is estimated (Guillet 1974) that a manufacturing plant producing 100,000 tonnes (10^8 kg) of synthetic fiber a year takes up an area approximately the size of a football field. In 2000, the world's production of synthetic textile fiber was 31.3 million tonnes (Anon. 2001), requiring by Guillet's estimate factories occupying a total of approximately 1 km² (about 250 acres or about 60 city blocks). Replacing this quantity of synthetic fiber with cotton would require about 500,000 km^2, or a land area roughly the size of France. For an equivalent amount of wool, grazing area for the sheep would need to be at least 12.5 million km^2, an area considerably larger than that of the entire United States (9.8 million km^2). So as the world's population grows and standards of living improve, it is clear that the Plastics Age is actually a necessity. Recall that in Chapter 1 we said that nature provided essentially no plastics. We will soon see that the search for a true plastic began about 150 years ago.

It will make sense to divide our summary of the history of polymers into three sections—before World War I, during the world wars, and after World War II—each reflecting the level of scientific and technical understanding in its period. It is not a coincidence that the events separating these periods are major military conflicts. Wars often separate nations from traditional natural resources, forcing the discovery of alternatives. Also, technological advances are often more numerous because governments seek advantages in weapons, communications, and transportation.

Polymers before World War I
Improving Natural Rubber

Humans have always modified objects of nature to make tools, clothing, and shelter. As the understanding of chemistry developed in the nineteenth century, it was common for scientists and inventors to chemically alter natural compounds and materials in an attempt to produce something more useful. Let's begin with natural rubber (NR), which, as we discovered in Chapter 3, occurs in certain plants and tropical trees. Spanish explorers recorded a game played with a rubber ball at the court of the Aztec emperor Montezuma II in the late fifteenth century. The shortcomings of NR were very well known—it is too soft and tacky on hot days and too brittle in the winter. For many years people had attempted any number of methods to modify its properties. In the early nineteenth century in Massachusetts, Charles Goodyear was obsessed with finding an additive that improved the properties of natural rubber (Kiefer 2002). Without any chemical or scientific background, he added one substance after another, achieving little success. In 1839, seriously in debt after at least five years' work, Goodyear accidentally spilled a rubber sample that contained elemental sulfur onto the stove. Instead of having to clean up a gooey mess, which NR by itself would have become, Goodyear found that the rubber/sulfur mixture was still flexible and remained so even at cold temperatures. Success at last!

Unable to generate interest or funding in the United States, Goodyear sent samples of his cured rubber to England. Thomas Hancock, a mechanical engineer with long-standing interest and experience in manufacturing rubber products, obtained some of them. Hancock developed his own sulfur/heating process and obtained a British patent in 1843, the year before

Goodyear obtained a U.S. patent. History is not clear on which inventor deserves the primary credit for this discovery. Regardless, Hancock named the process after Vulcan, the Roman god of fire and craftsmanship. *Vulcanization* transformed a readily available natural product with inferior properties into an extremely useful material. It is not a synthetic material, but rather a natural product made much more useful by altering its structure. Sulfur reacts with some of the double bonds in each chain, forming chemical crosslinks of one to about four sulfur atoms. Although the processing of natural rubber is today a complex science, vulcanization with sulfur is still part of the process. Vulcanization can also be carried out with many synthetic elastomers. Some 60 years or so after the invention, the automobile industry started its explosive growth, growth that would certainly not have been possible without the availability of vulcanized rubber for tires.

Behind the Eight Ball—The First True Plastic

In the 1840s many people were experimenting with cellulose. It was well known that cellulose reacts with strong mineral acids such as nitric or sulfuric acid. Christian Schönbein, a German who was professor of chemistry at the University of Basel in Switzerland, was very much interested in oxidation, a study that led him to discover ozone (O_3). He was aware that other chemists had reacted cellulose with nitric acid to form a derivative, *cellulose nitrate* (sometimes called *nitrocellulose)*. In 1845, Schönbein, in an apparent attempt to find additional substances that produce ozone, reacted cotton with a mixture of nitric and sulfuric acids. As we saw in Chapter 3, cotton is almost pure cellulose. In Equation 1 we show one repeating glucose unit in a cellulose chain. The squiggly lines indicate where the chain continues. Each repeating unit has three OH groups that can react with nitric acid to form nitrate groups. Using relatively less nitric acid produces cellulose nitrate with fewer nitrate groups and leaves a portion of the OH groups intact. Thus a family of cellulose nitrates exists. They differ in the fraction of OH groups reacted and differ in their properties.

Equation 1

Schönbein's product, which was almost completely nitrated as shown in the equation, had very unusual properties. When exposed to a flame, it exploded violently, but with extremely little smoke. Treating a sheet of paper (which contains a high percentage of cellulose) with his mix of acids made it transparent. Although the latter observation could have led him to a number of interesting materials and discoveries, Schönbein was far more interested in finding a buyer for his (almost) smokeless *guncotton*. He was convinced guncotton would replace conventional gunpowder, which on firing, produced huge volumes of smoke. He was unsuccessful in these efforts, in part because most of the governments he approached wanted to steal his idea rather than pay for it and also because laboratories or factories trying to produce it tended to explode unexpectedly. A useful explosive form of guncotton called *cordite* was not invented until 1889.

In 1846, the man who had first nitrated cellulose, Théophile-Jules Pelouze at the Collège de France, hired a young assistant named Louis Ménard to find solvents that would dissolve cellulose nitrate. The cellulose nitrate that Pelouze prepared was probably not as completely nitrated as Schönbein's. Ménard soon discovered that a mixture of ethyl ether and alcohol dissolved the cellulose derivative nicely to form a viscous, gel-like solution. The mixture, called *collodion*, could be applied like a varnish or lacquer to a surface, after which the solvents evaporated quickly, leaving a clear, tough layer. Collodion went on to many practical applications, including being an effective medical dressing for wounds, large and small. Perhaps its most significant use, however, was in photography as the medium in which bromide salts and silver nitrate were suspended, making an emulsion on glass plates. This so-called "wet plate collodion process" revolutionized photography and became the standard process for many years, replacing daguerreotype and other similar methods. Multiple positive prints could be obtained from one wet plate negative, meaning that for the first time photographs could be reproduced for mass enjoyment. However, the many attempts to produce solid objects from collodion all failed.

What drove the quest for a manmade material that could be shaped and molded into solid objects? In large part it was ivory, the hard, creamy white material harvested from the tusks of elephants and carved into attractive objectives. As the United States emerged from its Civil War in 1865, people resumed their normal lives and many desired possessions such as ivory-handled knives, ivory trays, combs and hairbrushes, ivory-backed mirrors, umbrellas with ivory handles, carved ivory jewelry, pianos with ivory keys, and, for the upper middle class, billiard balls for use on their new, popular pool tables. All of this demand was placing a huge toll on the elephant populations in such places as Ceylon (now Sri Lanka), Africa, and India. The best elephant tusks for making billiard balls came from the elephants of northern Ceylon, and at that, only about 2% of the tusks were of sufficient quality. At the time, no materials other than ivory were deemed suitable (Fenichell 1996).

Thus there was incentive to come up with a substitute material, a hard, smooth, substance with a density approximating that of the natural material. Preferably this substance would be able to be molded, thus allowing mass production of a myriad consumer items. They wanted a plastic! The solution

to making this plastic came after considerable experimentation by many en-
trepreneurs. Eventually it was discovered that mixing cellulose nitrate with
the natural product camphor produced a workable, moldable plastic that hardened on cooling. Depending upon the relative amounts of the two components, the end product, called *celluloid*, was hard and smooth like ivory, or
soft and pliable like leather. Although the American inventor John Hyatt is
usually credited with the invention in 1869, others (e.g., the Englishmen
Alexander Parkes and Daniel Spill) had also been hard at work trying to discover the magic compound (actually a *plasticizer*—see Chapter 7) that would
allow cellulose nitrate to be molded into solid objects. Hyatt also invented a
"stuffing machine" for forcing celluloid through a heated nozzle, allowing the
rapid mixing and manipulation (or *processing*) of the plastic as needed. His
machine was the precursor of the *injection-molding machine*, with which many
of today's plastic objects are made. Additional discoveries led to methods for
simulating the texture, color, and grain of native materials such as ebony, tor-
toiseshell, pearl, and marble. Suddenly the lower classes had access to personal and decorative items that resembled what until then only the wealthy
enjoyed, and a lucrative new industry was born.

The Picture of Things to Come

In the second half of the nineteenth century, photography was increasing in
popularity and importance as a chronicler of current events. Wet plates were
being replaced by dry-plate emulsions that could be prepared ahead of time,
dried, and then exposed later. No longer was it necessary for the photographer to carry along all of the chemical solutions, as well as the "dark" tent,
in which he or she disappeared to coat the plates, let them dry slightly, inserted them in the camera, and then emerged to take a picture. However, the
substrate on which the photographic emulsion was coated was still the rigid
and heavy glass plate. Cameras had to be loaded one plate (exposure) at a
time, and the plates had to be handled very carefully to avoid breakage.

George Eastman, who had opened the Eastman Dry Plate Company in
1881, and several other inventors recognized the advantages of being able to
roll up photographic emulsion on a flexible medium that someone had termed
film. Collodion was an obvious first choice for the substrate, but it was too
brittle and tore too easily. Just as with celluloid, the search was on for the
right additive(s) that would enable the formation of a freestanding, tough
plastic "film." Eventually Eastman developed such a formulation, a cellulose nitrate recipe similar to but more flexible than celluloid.

Eastman took the invention one step further by rolling a strip of the film
into a small, fixed-focus box camera he called the "Kodak," initiating the
birth of amateur photography. For the first time in history, individuals could
take spontaneous pictures of friends and loved ones or record images of
current events at will, a phenomenon that truly changed people's lives.

The availability of long, flexible strips of photographic film attracted the attention of other inventors. One was Thomas Edison, whose name is connected with many inventions and innovations (e.g., the phonograph, the incandescent lightbulb, the telegraph, electricity distribution). Over the last two decades of the nineteenth century, Edison was interested in "doing for the eye what the phonograph does for the ear" (Fenichell 1996). After a few false starts, Edison and his coworkers decided that, by punching holes along the edges of celluloid photographic film (called perforations), they could produce "motioned pictures." In France, the Lumière family was after the same objective, and produced the "cinématographe," a motion-picture projection system. For the first time, people could view actual events, "filmed" as they happened. The result was astounding. The average citizen could view a historic event such as the coronation of a king or a czar to which only the wealthy and well connected had previously had access. And the average citizen could be entertained *visually*, rather than by just listening to the radio. Ultimately what were needed were long rolls of cellulose nitrate photographic film, which George Eastman was most willing to provide. So once again, celluloid, the first plastic, changed the lives of the common citizen in very dramatic and far-reaching ways.

Recall that another use of cellulose nitrate was as guncotton, the smokeless gunpowder. Clearly a concern about using this cellulose derivative was

George Eastman, left, holds Thomas Edison's "motioned pictures" film while Edison studies the camera. Courtesy George Eastman House.

its flammability. Until it was replaced in the 1940s by another cellulose compound, cellulose acetate, fires occurred periodically, resulting in the destruction of a few theaters and in a number of deaths.

Other Cellulose Derivatives

The motivation to make derivatives of cellulose was the conversion of the intractable material into something that could be dissolved and then processed. As we just saw, *cellulose acetate* replaced cellulose nitrate in commercial motion picture film and was the only polymer ever used for home movie film. Edison introduced home moviemaking in this country in 1911. The acetate derivative is prepared by treating cellulose with sulfuric acid and acetic anhydride, a reactive compound derived from acetic acid. See Equation 2. (For simplicity, we show only one OH group bonded to a repeat unit of cellulose.)

Equation 2

$$\text{cellulose}{-}\text{OH} \ + \ CH_3\overset{O}{\overset{\|}{C}}{-}O{-}\overset{O}{\overset{\|}{C}}{-}CH_3 \ \xrightarrow{H_2SO_4} \ \text{cellulose}{-}O{-}\overset{O}{\overset{\|}{C}}{-}CH_3 \ + \ CH_3\overset{O}{\overset{\|}{C}}{-}OH$$

Cellulose acetate forms strong, transparent films and has enjoyed many applications such as photographic film, transparent tape, and blister packaging. It can also be spun into satin fibers. *Satin* not only means a somewhat shiny fabric woven from cellulose acetate fiber, but also refers to something with a soft texture, which the acetate fiber has.

An ingenious treatment of cellulose was discovered by Charles Cross and Edward Bevan in England in 1892. It involved first preparing a chemical derivative called *cellulose xanthate* in a process that is conceptually no different from converting cellulose into other derivatives such as cellulose acetate or cellulose nitrate. What made this different, however, is that xanthates are reactive chemical intermediates that can be converted easily into still different compounds, or returned to the starting material, in this case cellulose. See Equation 3.

Equation 3

$$\text{cellulose}{-}\text{OH} \ + CS_2 + \ NaOH \ \longrightarrow \ \text{cellulose}{-}O{-}\overset{S}{\overset{\|}{C}}{-}S^{-} Na^{+} \ \xrightarrow{H^{+}} \ \text{cellulose}{-}\text{OH}$$

$$\text{xanthate} \qquad\qquad \text{regenerated cellulose}$$

$$+ \ H_2O \qquad\qquad\qquad + \ CS_2 \ + \ Na^{+}$$

Cellulose xanthate is soluble in aqueous base, the solution being called *viscose*. This can then be forced or extruded through very small holes (called *spinnerets*) into a sulfuric acid solution, which removes the xanthate groups and regenerates cellulose. But when the cellulose reforms, it is in long continuous fibers called *viscose rayon*, originally referred to as artificial silk and popular for a long time as a fabric for clothing. Alternatively, the viscose solution can be extruded through a slit, forming a *film* of cellulose, called *cellophane*. This clear, tough material was used for many years, as cellophane tape and as a film for wrapping any number of consumer items, including fresh produce, meat, candy, cigars and cigarettes, soaps, and perfumes. It was also found to make an excellent semipermeable membrane for kidney dialysis, as well as an effective, and transparent, dressing for wounds. Another important development was the regeneration of cellulose in a three-dimensional form filled with tiny holes: a synthetic sponge. A sponge is a primitive marine animal (phylum *Porifera*) with a porous fibrous skeleton that has been harvested and used in many parts of the world as a "sponge." As the world's population grew, the price of native sponge also grew. Therefore a market existed for a synthetic version, and most of the "sponges" one finds in stores today are actually regenerated cellulose. The material is *hydrophilic* (it is attracted to water), yet it does not dissolve.

Viscose rayon is but one variety of rayon, a more general term for derivatized or reconstituted cellulose. Other rayons include fiber prepared from collodion, cellulose acetate, and cellulose fiber regenerated from a cellulose–copper ammonium solution (*cuprammonium rayon*) (Kauffman 1993).

We noted at the start of this chapter the tremendous importance of cellulose in human existence. How interesting that we have been trying for at least a century and a half to improve it. And we have done that.

The First Synthetic Polymer

So far we have discussed only products that result from the modification of existing polymers such as natural rubber or cellulose. What was the first "real" synthetic polymer, a material polymerized totally from small-molecule organic compounds? In 1902, a young Belgian chemist working in Yonkers, New York, by the name of Leo Baekeland was attempting to do just that. He had made a small fortune selling his fast photographic paper to George Eastman a few years earlier.

There was a growing need for a material that could be used for electrical insulators as an increasing number of homes and businesses obtained electricity. Several scientists had experimented with reactions between the two organic compounds *phenol* and *formaldehyde*. Baekeland knew that the products of this reaction were almost always insoluble, brittle masses. Because the reaction tends to give off gases, he finally decided to carry out the process in what amounted to a pressure cooker, thereby preventing the gases from escaping. His instinct was correct, and he worked out conditions to obtain a liquid called *Bakelite resin* that hardened into a yellow-red, transparent solid in the shape of the container in which it had formed. The solid was insoluble in every solvent to which it was subjected, and it did not soften on heating. He had invented the first synthetic thermoset resin. See Equation 4. By contrast, celluloid is a thermoplastic, a substance that can be softened on reheating and can be dissolved.

Equation 4

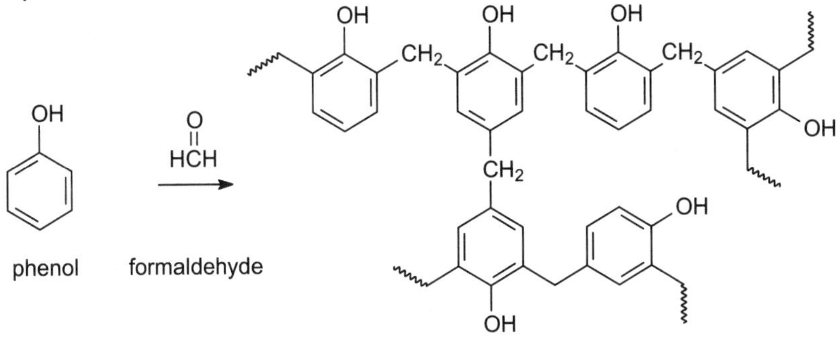

phenol   formaldehyde

Not only was there a large market for Bakelite for electrical insulators, but it also became the material of choice for manufacturing a wide range of consumer items, including billiard balls. It was the "material of a thousand uses" (Fenichell 1996). Because it was inherently strong and would not burn or soften, it was perfect for electrical fixtures, telephones, radios, radiator caps, toothbrushes, eyeglass frames, fountain pens, cameras, and ashtrays, to name a few. Fashion designers took to Bakelite enthusiastically as the Art Deco movement grew in the late 1920s, because it could be molded into appliances and furniture of almost any shape. Until replacement materials were developed years later that could be produced in a range of colors, Bakelite was virtually the only synthetic thermoset resin available. It could be formulated with wood or other two-dimensional materials to make laminated countertops. The liquid could be applied to surfaces such as wood and then hardened into an extremely durable coating.

Baekeland succeeded where others had failed, not only because he carried out his process under pressure but also because he recognized the importance of controlling the relative proportions of phenol and formaldehyde. Too little formaldehyde, for example, and the product was weak and brittle. Baekeland also worked out conditions for reacting phenol with a small amount of formaldehyde under acidic conditions, to produce a thermoplastic material he called *novolac*. Novolac can be converted into a thermoset similar to Bakelite by heating with additional formaldehyde. Novolac became useful as a positive-working photoresist for the manufacture of integrated circuits some 60 or 70 years later. We will discuss photoresists later in this chapter.

Polymer Chemistry Defined by the World Wars
What Is a Polymer, Anyway? Call in the Special Forces

So far we have seen the development of a number of polymers and somehow assumed that those who produced them knew what they were working with. Not exactly. Actually, up through the first quarter of the twentieth century, no one really knew what a polymer was. Two schools of thought had developed on the fundamental nature of polymers. One group of leading scientists was thoroughly convinced that atoms could bond together to make molecules of up to only a certain size. Substances larger than this critical size, the argument went, were made up of collections of smaller molecules called *colloids*. The molecules making up a colloid were held together by special forces called "secondary valence forces." The scientists cited considerable experimental evidence that was consistent with the known behavior of natural polymers such as proteins or polysaccharides, thereby supporting this hypothesis. A few, however, did not accept the popular opinion that molecules were limited to a certain, finite size. In addition, they questioned the necessity of invoking the hypothesis of secondary forces to explain the experimental data.

Hermann Staudinger and Herman Mark: Who's the Father?

Just as in many other fields of endeavor, significant progress in polymer science often depends upon large, bold steps by a few courageous, independent-thinking pioneers. These are people who can set aside the current thinking and provide their own analysis. They look at the same things we do, but see something quite different. They often ask fundamentally different questions, and, as a result, their answers lead them down different paths. For this insight, they are often scoffed at and their conclusions rejected by those in the mainstream. In 1921, Hermann Staudinger (1881–1965) was a well-established, respected German professor of organic chemistry with a prestigious chair at the federal Institute of Technology in Zurich, Switzerland. He was 40 years old, a mature scientist, and author of a number of significant publications in organic chemistry. He was destined to become one of the most respected leaders of organic chemistry. He decided to turn his attention to polymers or "high molecular compounds," at that time not even what one could call a field of chemistry. For this he was openly ridiculed and attacked by other organic chemists, who viewed polymers as "very unpleasant and poorly defined" materials unworthy of serious study (Staudinger 1970).

Fortunately, he persisted, made a number of very significant discoveries, and was responsible for helping put polymer chemistry on a sound chemical foundation. Many of his students went on to become academic and industrial leaders in polymer science. He applied his organic chemistry insights to the study of what were finally accepted as very large organic molecules. For his many contributions, he received the Nobel Prize in chemistry in 1953. He has frequently been called the "father of polymer chemistry."

But what about that other Herman? Herman Mark (1895–1992) was an Austrian whose academic career began in Berlin. While there he studied natural polymers such as cellulose, silk, and wool using X-ray crystallography. By the mid-1920s he was presenting experimental evidence in support of Staudinger's arguments for the existence of macromolecules, thereby helping settle the controversy. He soon left the university for industry, becoming assistant research director at the huge chemical company I. G. Farben, where he was in charge of research on plastics. He developed a practical synthesis of styrene, which allowed the production of inexpensive polystyrene, and continued to study the physical properties of polymers. Forced to leave first Germany, then his native Austria by the Nazis in the 1930s, Mark eventually ended up at the Brooklyn Polytechnic Institute in New York, where he founded the Polymer Research Institute, the first of its kind in the New World. His contributions to polymer science are many and include both experiment and theory. His description of the solution behavior of polymers (see Chapter 6) resulted in a useful relation between viscosity and molar mass that still bears his name (the Mark-Houwink-Sakurada equation). Mark brought knowledge and insights of the developing field of polymers in Europe to scientists in the United States and Canada. He started the first American undergraduate polymer courses and spoke tirelessly on polymer science to widely diverse audiences, including early television viewers. Like Staudinger, Mark mentored a number of scientists who went on to make very significant contributions in the field. He continued to write and lecture well into his nineties (Morris 1986). This Herman is appropriately called the "father of polymer physics."

Wallace Carothers and the First Synthetic Fiber

In the late 1920s, a young Wallace Carothers left his teaching position at Harvard to help start a new basic research group in organic chemistry at the DuPont Experimental Station in Delaware. Carothers, called "the best organic chemist in the country" by his mentor Roger Adams at the University of Illinois, was an organic chemist with a keen interest in the nature of chemical bonding (Raber 2001). He was convinced that large molecular chains could be constructed using the proper chemistry and set out to do just that. DuPont promised him a laboratory, a small group of bright young chemists, and the freedom to explore whatever avenues he wished.

Carothers was drawn to the controversial papers of Staudinger, who argued that polymers were generally made up of long chains of atoms bonded together, just like the bonds in small molecules. Fluent in German, Carothers could understand Staudinger's complex arguments although many Americans could not. Carothers reasoned that he should be able to design the laboratory synthesis of macromolecules with structures similar to those of natural polymers by building the chains from small units. This is significant because Carothers was really the first chemist to approach the synthesis of polymers rationally. At that time, the largest molecule ever synthesized had been reported by Emil Fischer, the great nineteenth century organic chemist, and had a molar mass of 4200. Carothers wanted to beat this record. He also argued that, just like the natural polymers, his synthetic compounds should have useful properties.

Being an organic chemist, Carothers reasoned that the right condensation reaction should in principle lead to a long-chain polymer. It was well known that a carboxylic acid and an alcohol can react together to form an ester plus water:

Equation 5

$$R-\overset{\overset{\displaystyle O}{\|}}{C}-OH \quad + \quad R'-OH \quad \rightleftharpoons \quad +R-\overset{\overset{\displaystyle O}{\|}}{C}-O-R' \qquad H_2O$$

So Carothers figured that a *di*-carboxylic acid and a *di*-alcohol could react to form a *poly*ester:

Equation 6

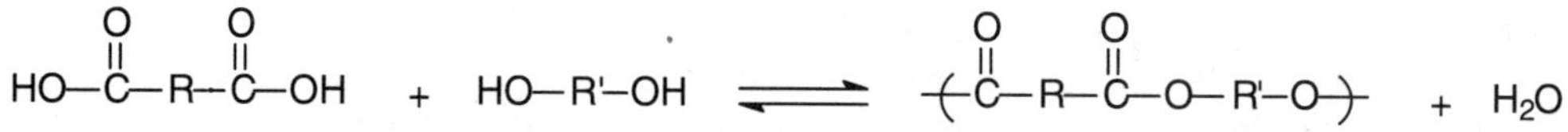

$$HO-\overset{\overset{\displaystyle O}{\|}}{C}-R-\overset{\overset{\displaystyle O}{\|}}{C}-OH \quad + \quad HO-R'-OH \quad \rightleftharpoons \quad \left(\overset{\overset{\displaystyle O}{\|}}{C}-R-\overset{\overset{\displaystyle O}{\|}}{C}-O-R'-O\right) \quad + \quad H_2O$$

For over a year Carothers and his group made polyesters in their laboratory, each time obtaining product with a molar mass of 3000 to 4000. As noted in the equations above, the esterification reaction is reversible. Ester groups can be hydrolyzed (react with water), to reform carboxylic acid and alcohol. When they worked out a procedure to remove the last traces of water, Carothers and his team obtained polyester with a molar mass of approximately 12,000. The value was far above that needed for molecular entanglement, a significant achievement. They were elated.

From Globs to Fiber

When Carothers's assistant, J. W. Hill, tried to remove the polyester from the reaction vessel, the melted mass came away in long fibers. He was able to make the fibers even longer by stretching them when they had cooled (*cold drawing*). Amazingly, the fibers became *stronger* as they got thinner and longer. They were transparent, very elastic yet tough, and could be tied into tight knots without breaking. Their strength was similar to that of silk and cotton (Morawetz 1985). As we will discuss further in Chapter 7, stretching a fiber causes alignment of the molecules in the fiber. The increased strength arises from the increased molecular order in the fiber. This was a very exciting time in Carothers's DuPont laboratory.

Subsequent work led the team to the conclusion that, while they had figured out the tricks to make very high molar mass polymer, beating Emil Fischer's record several times over, the ultimate properties of the polyesters they were making were not good enough for useful products. These polyester fibers dissolved too readily in dry-cleaning fluid and softened in hot water. Imagine trying to clean a shirt prepared from this polyester. Polyesters with different structures would be needed before any kind of commercialization could be contemplated. Carothers had been successful in figuring out how to synthesize polyesters but had made a compound with little commercial potential.

Rather than pursuing different polyesters, for some reason Carothers turned his attention to polyamides, polymers with the same functional group as in proteins. He figured, correctly, that crystallinity in the polymer as well as hydrogen bonding between chains would greatly improve the physical and mechanical properties of the fiber. This work led to the development of an exciting new family of synthetic polymers that DuPont named *nylon*. The premier polymer was nylon-6,6, or poly(hexamethylene adipamide) (see Equation 2, Chapter 2).

Carothers had favored nylon-5,10 which is prepared from a five-carbon diamine and a 10-carbon diacid rather than nylon-6,6. However, Elmer Bolton, his practical, profit-minded boss pointed out that the starting materials for nylon 5,10 would be expensive on a large scale. For example, the diacid, sebacic acid, came from castor oil, which is obtained in limited quantities from the castor bean. There would not be enough castor oil to provide sufficient sebacic acid for the amount of nylon that he envisioned. Instead, Bolton reasoned that only one petrochemical, benzene (C_6H_6), could provide both of the starting materials for nylon-6,6. This turned out to be a very shrewd decision. And so this became the world's first truly synthetic commercial fiber.

The marketers at DuPont announced their new wonder fiber to the public at the 1939 New York World's Fair. Amid a huge display showing off a wide range of products made of plastic, they featured nylon fishing line, nylon toothbrushes, and of course, nylon stockings. At the time, women wore stockings made only of silk, which was both expensive and exceedingly fragile (see Chapter 3). At the time the *sheer* look was "in," so stockings were very thin and almost transparent. Usually, after only one wearing, they had to be thrown away. DuPont took every opportunity to hype more durable stockings made of their new synthetic fiber, for a while selling only one pair per customer in a limited number of stores. Women lined up hours early to make sure that they could purchase a pair before the store sold out. And they were willing to pay *more* for "nylons" than they would for a pair made of silk ($1.15 to $1.35 versus $0.79 for silk). This was the first time in history that a synthetic material was judged more valuable than the natural material it had replaced (Fenichell 1996).

It is interesting to see what else was featured at the 1939 World's Fair. Plastics were "in," and both DuPont and Rohm and Haas showed off many household items made of transparent poly(methyl methacrylate) (Lucite and Plexiglas, respectively). In addition, there was safety glass—two sheets of glass sandwiching a layer of soft poly(vinyl butyral), a transparent plastic house, and a Plexiglas Pontiac automobile. It was also the first time television was seen by the public, who stood in long lines to get a glimpse of a set whose case was made out of Plexiglas.

Within a few years, a number of other fibers became commercially available, including useful polyesters, a class of compounds that Carothers had vacated in favor of the polyamides.

Chain-Growth Polymers

As Carothers and others concerned themselves with making high molar mass step-growth polymers, several groups of chemists were experimenting with what we now call chain-growth polymers. Staudinger and his colleagues had been studying compounds that contain a carbon-carbon double bond such as styrene for several years. These are referred to as *vinyl* or *addition monomers*.

By 1929, when he published a key paper, Staudinger had figured out that these monomers polymerized by a rapid chain reaction involving some kind of *activated monomer* or *reactive intermediate*. Monomer molecules can add to this species, forming a new reactive intermediate at the end of the growing chains until some termination process occurs. He and others knew that peroxides and sunlight catalyzed these polymerizations. These fundamental studies attempting to understand the nature and mechanism of vinyl polymerizations attracted the attention of a wider audience, including industrial chemists.

This included the German chemical giant IG Farben, which was staffing its research laboratory at approximately the same time that DuPont was beginning its basic research laboratory. (Herman Mark had joined Farben in 1927.) Farben patented an adhesive based on poly(vinyl acetate) that was polymerized in the presence of sunlight. One of the scientists suggested that the "poor light intensity during the winter months may be taken care of by increased production from March to October" (Morawetz 1985).

Increased commercial interest in vinyl polymers catalyzed an increasing number of studies on how monomers reacted as well as on better understanding the properties of the polymers. The result was overwhelming support of Staudinger's macromolecule hypothesis and an eventual fading of the colloidal aggregation school. Staudinger continued his careful, pioneering studies, and was awarded the Nobel Prize in chemistry in 1953 in recognition of the importance of his work.

Paul Flory, whom Carothers hired at DuPont in 1934, was a physical chemist who devoted much of his career to polymer physical chemistry. Over the course of his career, Flory made a huge number of significant contributions to the understanding of polymerization and polymer physical properties. For these efforts he was awarded the Nobel Prize in chemistry in 1974.

Polyethylene and the War

At the same time in England, Imperial Chemical Industries (ICI) was also hiring chemists to carry out basic research. E. W. Fawcett and R. O. Gibson there worked out conditions to prepare polyethylene using high-pressure reactors, which unexpectedly tended to explode on occasion. After one explosion that caused considerable damage to a laboratory, they were told by management not to do any more work on the polymer. Two years later, their immediate supervisor, M. W. Perrin, secretly tried one last experiment, suspecting that the polymer might actually be useful. The polymer they synthesized, for which they received a patent in 1939, is a white, waxy solid with unremarkable physical properties. However, it is an excellent electrical insulator, a property that enabled the British to develop a radar system. With radar, the British could detect approaching enemy aircraft, even at night or under cloudy conditions, an advantage that allowed the heavily outnumbered Royal Air Force to defeat the German Luftwaffe in the Battle of Britain. Today polyethylene is our largest commercial polymer. Specialized catalysts and reaction conditions are used to produce a family of polyethylene polymers having a range of chain architectures and properties. We will discuss this more thoroughly in Chapter 5.

Figure 4-1. Natural rubber prices in the United States during first third of the twentieth century.

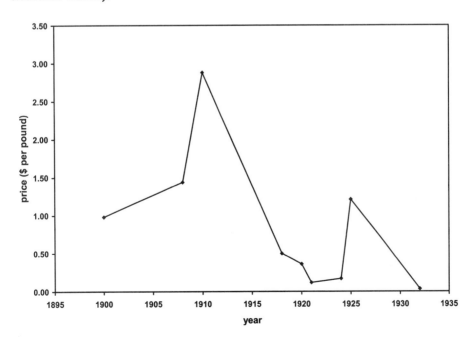

It's a Stretch

As we saw, the availability of a steady supply of natural rubber was essential as the automobile became popular in industrial countries. During times of war, it was possible that one country or another's access to rubber would be cut off. Thus there were times when work on developing synthetic rubber was quite urgent. In addition, the price of natural rubber varied wildly for the first third of the twentieth century, shown vividly in Figure 4-1. A blockade by the British during the First World War forced Germany to produce a serviceable synthetic rubber, which, however, required a reaction time of several months! In the mid 1920s, the price of natural rubber was quite high once again, causing chemical companies in several countries to reinitiate work on synthetic rubber. In 1927 in the United States, J. C. Patrick accidentally discovered Thiokol when he reacted 1,2-dichloroethane with sodium polysulfide:

Equation 7

$$Cl-CH_2CH_2-Cl \ + \ Na_2S_x \ \longrightarrow \ (CH_2CH_2-S_x)_n \ + \ 2\ NaCl$$

In the equation, x is a small integer, usually from 2 to 4. Thiokol became the first synthetic rubber manufactured in the United States. The sulfur atoms in the polymer backbone offer protection against the elements and limit the solubility of the polymer in organic solvents. As a result, Thiokols have excellent resistance to gas and oil and are still used for seals, gaskets, O-rings, and hoses in parts exposed to petroleum products. They are also used as binders for solid rocket propellants. (Footnote: We should point out here that carbon and sulfur are the only two elements in the periodic table that bond to themselves and form long chains. Elemental [rhombic] sulfur, which normally exists as cyclic molecules with eight sulfur atoms in a ring, can be polymerized into long chains on heating. However, the polymer has very limited practical usefulness because the chains depolymerize and reform rings over time.)

Because it was known that isoprene was the building block of natural rubber, many people investigated its polymerization in addition to that of various other *dienes*, hydrocarbon monomers containing two double bonds. In Germany, chemists discovered that 1,3-butadiene could be polymerized with sodium metal to make a useful synthetic rubber called *Buna*. The name comes from *bu*tadiene and *na*trium, the Latin name for sodium. Other materials soon followed, including Buna-S (a copolymer of butadiene and styrene):

Equation 8

$$CH_2{=}CH{-}CH{=}CH_2 \ \xrightarrow{\ Na\ } \ -(CH_2{-}CH{=}CH{-}CH_2)_n{-} \qquad -(CH_2{-}CH{=}CH{-}CH_2)_m{-}(CH_2{-}CH)_n{-}$$

butadiene poly(1,4-butadiene) [Buna] Buna-S

At about the same time, Wallace Carothers and his colleagues discovered a chlorine derivative of butadiene that polymerized to give a rubbery polymer that DuPont eventually marketed as Neoprene. Note the similarity in the structures of isoprene and 2-chloro-1,3-butadiene (chloroprene):

$$CH_2{=}C{-}CH{=}CH_2 \qquad\qquad CH_2{=}C{-}CH{=}CH_2$$
$$\qquad\; CH_3 \qquad\qquad\qquad\qquad\quad Cl$$

isoprene   chloroprene

Polychloroprene (Neoprene) is similar to natural rubber, except that the synthetic material has better resistance to ozone, sunlight, and many chemicals.

World War II brought a renewed urgency for synthetic rubber. When the Japanese occupied the Malay Peninsula and adjacent islands, the primary source of natural rubber to the United States was cut off. Rubber companies and university scientists cooperated in developing a procedure for manufacturing a synthetic rubber called *GR-S* (Government Rubber—Styrene). By the late 1930s, American scientists had learned a little about Buna and Buna-S that the Germans had developed. Building on this rudimentary knowledge, a procedure was worked out to produce GR-S by emulsion polymerization (see Chapter 5). Rather than an organic solvent, water is used as the medium to disperse the monomers and polymers and to remove heat. Emulsion polymerizations proceed rapidly (thus allowing high productivity) and produce polymers with high molar mass. Within a matter of months, production of GR-S had begun and new government-owned plants were under construction. As a result of this effort, the United States could maintain a supply of tires and other rubber parts for airplanes, trucks, and jeeps, a key factor in helping the Allies win the war.

Down the Slippery Slope

How many times have we mentioned that one polymer or another was discovered as the result of a lucky accident? Here's one more. In 1936, Roy Plunkett (1911–94) had finished his doctoral work in chemistry at Ohio State University (where he overlapped with Paul Flory) and had taken a job at DuPont working on halogen-containing compounds. DuPont manufactured chlorofluorocarbons (CFCs), compounds of carbon, hydrogen, chlorine, and fluorine. These compounds, called *freons* and once commonly used in refrigerators and air conditioners, are atmospheric ozone depleters and have now been replaced in most parts of the world with compounds that contain no chlorine. At one point, Plunkett prepared a large quantity of tetrafluoroethylene gas ($F_2C{=}CF_2$), which he transferred to metal cylinders for storage. When the valves to the cylinders were opened the following day, nothing came out, although no gas had leaked out because the tanks had not lost any mass. Curious, he sawed the cylinders in half, finding a white powder inside coating the walls of the tanks.

The white polymer was very unusual. It would not melt, even when subjected to very high temperatures, and did not burn. It did not dissolve in any organic solvent, nor in extremely strong acids or bases. Besides that, it was very slippery and nothing would stick to it. Plunkett knew that the gas had spontaneously polymerized to form polytetrafluoroethylene, later to be called Teflon. No one had a clue what to do with it.

Enter the United States Army. In 1942, Lieutenant General Leslie Groves visited DuPont in search of help with the Manhattan Project, the top-secret effort to build the atomic bomb. Separating uranium-235 for nuclear weapons from the much more plentiful uranium-238 required the generation of UF_6, an extremely corrosive gas. An inert material was needed for gaskets. The polytetrafluoroethylene discovered just four years prior was the perfect (if expensive) answer. The polymer found a number of other applications during the war as well, elevating it from a laboratory curiosity to one of our most unusual, but extremely useful polymers. It was found to be even a better insulator than polyethylene and so was also used in the development of radar in the United States. Several years later, grateful consumers were pleased to own nonstick Teflon-coated frying pans and bakeware.

The period from World War I to the end of World War II saw the introduction of a number of polymers with useful properties that began to change our lives forever. Some materials, such as nylon, replaced more expensive natural fibers such as silk. Polymers such as synthetic rubber enabled countries separated from traditional natural resources to continue to wage war or to defend themselves. And still other polymers (e.g., polyethylene and polytetrafluoroethylene) enabled the development of entirely new technologies and industries (e.g., radar and telecommunications).

After the Second World War
A Whole New Approach to Chain-Growth Polymers

By the early 1950s, many countries were experiencing growing economies. A number of new consumer goods, such as televisions, new automobiles, and kitchen gadgets, were now widely available, and the public was eager to buy. Some of the items for sale were constructed of a new material called plastic. Chemical companies were investing large sums of capital in new products, including polymers. As a result, a number of new polymer products had recently been introduced, including polystyrene (a rigid and transparent polymer), high-pressure polyethylene (soft, tough, and flexible), and poly(vinyl chloride) (a versatile plastic that is also fire-resistant). All of these materials were relatively inexpensive and easy to fabricate into useful products.

As we saw above, polyethylene had been commercialized a decade earlier and was produced by polymerizing ethylene in a high-pressure, high-temperature reactor. Because of the high temperature and the mechanism by which this polymerization proceeds (see Chapter 5), the polyethylene chains contain a large number of branches. Because of these branches, the chains do not pack very efficiently, giving the polymer sample a relatively low density. Thus the name low-density polyethylene (LDPE). If one could make perfectly linear, unbranched polyethylene, the polymer would probably have somewhat different and potentially useful properties.

As luck would have it (literally), such a polymer was soon discovered. By 1953, Dr. Karl Ziegler was a very well respected organic chemist and director of the prestigious Max Planck Institute for Coal Research in Mülheim, Germany. He had spent many years investigating metal-organic compounds, compounds of carbon, hydrogen, and so on, that also contain a metal atom. Because carbon-metal bonds are not very stable, these compounds are very reactive and are often used as intermediates in the preparation of new organic compounds. In 1953 Ziegler and his associates were working with the reactions of olefins (hydrocarbons with carbon-carbon double bonds) with organoaluminum compounds. Because many of the olefins, including ethylene, are either gases or low-boiling liquids, the reactions were carried out in pressurized reactors. A reaction in only one of Ziegler's several pressure reactors had produced an unexpected product. Extremely careful analysis suggested that the reactor had become contaminated with a trace of nickel.

Ziegler, an organic chemist who had little experience with polymers, recognized that this unanticipated result might lead to a method to make linear polyethylene. He knew that there would be a demand for such a material from any number of industrial companies, and he wanted to be the one to hold the first patents. He told his student, Heinz Breil, to test the reaction with ethylene systematically with as many other metals as he could find. In a short time, after trying a number of different metals, Breil ran a reaction laced with zirconium and produced a large quantity of linear polyethylene. The properties of this polymer were superior to those of any polyethylene known to date. Without branching, sections of the chains tend to line up together, resulting in a polymer with a relatively high density (HDPE). The new polymer was considerably harder than LDPE, and tougher. Ziegler and his coworkers knew at once that they had achieved a very significant breakthrough.

From here the story takes some interesting turns. Ziegler was very much in charge of his institute, personally directing the research of every senior scientist, postdoctoral scientist, and student. No detail escaped his attention. In addition, he traveled often, giving lectures throughout Europe and sometimes in the United States. He welcomed an impressive number of visitors to his institute, scientists from both universities and industry. He personally drafted licensing agreements with various chemical companies, allowing them, for a fee, the right to develop the chemistry covered by the agreement. Many of these agreements pertained to new metal-organic compounds and their reactions. A few companies, looking for new products to add to their arsenals, were interested in polyethylene. They would pay dearly to have access to a new polyethylene with superior properties. And several did.

How the Hula Hoop Saved the Day

Owning the recipe to a polymerization reaction on the laboratory scale along with having the right to use it is one thing. Figuring out how to produce high quality polymer in an industrial scale process by the tonne is something entirely different. Essentially Ziegler was selling just the rights to his recipe. The path to the sale of products molded from HDPE turned out to be quite tortuous. In the early 1950s in the United States, Paul Hogan and his colleagues at the Phillips Petroleum Company in Oklahoma had also developed catalysts that produced linear polyethylene. Over the next few years, Phillips engineers worked out a process for manufacturing the new polymers and were selling the rights to their process to other companies.

Thus, several years after the discoveries of linear polyethylene, a number of companies both in the United States and in Europe had developed the process and began selling products such as blow-molded bottles and extruded sheet and pipes. Pipes made of HDPE promised to replace those of much more expensive steel and copper for plumbing applications. However, just as some of these plants were beginning to go to full production, disaster struck. The bottles developed cracks and the pipes, in time, began to bulge and then leak. How could such a strong polymer fail so miserably? And more important, how could these plants, which had just consumed millions of dollars in construction and development costs, keep on going? It was clear to all that considerable time would be needed to unearth the source of the problem and to come up with a remedy. As luck would have it (again!), rescue was at hand and in the least likely of places—a plastic toy!

In the spring of 1957, the Wham-O Company of California was looking for a new toy that would bolster its sales. You may have heard of them—in more recent times, Wham-O has brought us the Superball and the Hacky Sack. At the time, Wham-O was best known for its wooden slingshots, something no young boy could be without. Arthur "Spud" Melin and his partner Richard Knerr heard about bamboo hoops approximately three feet in diameter that kids in Australia were crazy about. Eureka! Wanting something more colorful than dull old bamboo, Melin and Knerr experimented with a new plastic just on the market that came in many bright colors—HDPE. They could manufacture their product from extruded tubing for approximately 50¢ each. They called it the Hula Hoop, and a fad began. The popularity moved from west to east, and by the spring of 1958, sales exceeded 15 million at about $2 each. At the peak of its popularity, they sold 25 million Hula Hoops in four months. This created a demand that forced the new polyethylene manufacturing facilities into round-the-clock operation. This unlikely, almost unbelievable market generated the cash and bought the time necessary for the polymer chemists and engineers to figure out how to fix the somehow flawed HDPE for more demanding applications. We will discuss the molecular nature of the problem and its solution in Chapter 7. Solving the problems enabled HDPE to go on to become one of the largest volume commodity polymers. In the meantime, Melin and Knerr had plans for the day when the Hula Hoop craze died. They called it the Pluto Platter, better known now as the Frisbee.

From Polyethylene to Polypropylene—Stereoregularity

Not surprisingly, the eventual success in the new polyethylene made Herr Doktor Ziegler a very wealthy man, as well as a Nobel laureate, a prize he shared with Giulio Natta in 1963. As Ziegler's work with linear polyethylene was progressing, the question arose as to what other olefins might polymerize with his catalysts. The next in the series would be propylene, three carbons instead of two for ethylene. As we said, it is apparent that Ziegler made most of the decisions in his institute. Among the throng of visitors that Ziegler entertained during the early 1950s were scientists from Professor Giulio Natta's group at the Milan Polytechnic Institute in Italy. Unlike Ziegler, Natta was a polymer chemist and engineer. In fact, he was first introduced to polymers by none other than Hermann Staudinger. But like Ziegler, he balanced his interest in and understanding of chemistry with a desire to produce something practical. He sought to solve the problems of industry using the scientific method (McMillan 1979). And like Ziegler's, his institute was heavily funded by industry, in this case Montecatini, Italy's largest chemical company (later Montedison).

When Natta heard Ziegler lecture on his metal-organic catalysts in 1952, he, like few others, understood at once the potential significance. He convinced officials at Montecatini to sign a licensing agreement with Ziegler that, among other things, allowed Natta to share information with Ziegler. Thus it was that some of Natta's/Montecatini's scientists were working in Ziegler's institute in 1953.

The story of the interactions of these two "giants of polyolefins" is far too long and complex to repeat here. The reader is directed to Frank McMillan's *The Chain Straighteners* (1979) for an excellent and full accounting of this fascinating history. In short, following Ziegler's success with polyethylene, his friend Natta asked him if the catalysts would also polymerize propylene. The only polypropylene known at the time was an oily liquid. Polypropylene from Ziegler's process might have altogether different properties. In response to Natta's question, Ziegler is quoted as saying emphatically "Es geht nicht" (It does not work). Actually at this point, Natta had already successfully polymerized propylene with Ziegler's catalyst, but kept the information to himself, essentially undermining the agreements he had negotiated with Ziegler.

Natta lost no time patenting his process, thus assuring himself and his backers of fame and fortune. Because of his background in polymer chemistry, Natta was able to analyze his new product thoroughly and to understand the molecular basis of its properties. His polypropylene was a highly crystalline polymer, the crystallinity enhanced by the fact that all of the CH_3 groups lined up on the same side of the polymer chain. We call polymers such as this *stereo-regular*. (We will discuss this in Chapter 5 in the section "Polymer Stereochemistry: Tacticity"). All of this was quite new at the time, although Paul Flory had anticipated the possibility in his new, now-classic book *Principles of Polymer Chemistry* published just a few months earlier (1953).

As is often the case, a number of other people had been working with catalysts similar to Ziegler's during these years in the 1950s. Some of them succeeded in making linear polyethylene or polypropylene, but failed to capitalize on it. Some came very close but, for a variety of reasons, did not quite succeed. As we mentioned, Ziegler and Natta shared the Nobel Prize in chemistry in 1963. Although their friendship had long since dissolved, in public, at least, they showed respect for each other. Early on, the catalyst systems became known as Ziegler catalysts, the name bestowed by none other than Natta. We now refer to this general class of polymerization as Ziegler-Natta polymerization. Several other monomers have been polymerized by this route, including a number of dienes, producing the first synthetic polymer with the same structure as natural rubber (cis-1,4-polyisoprene). Subsequent development led to several useful synthetic rubbers such as polybutadienes and ethylene-propylene-diene copolymers (EPDM). The search for, and design of, other metal-based catalysts that control stereochemistry is still a very active field and has resulted in the development of new families of polymers. We will investigate this further in Chapter 5.

Other Significant Advances
Chain-Growth Copolymers
Random Copolymers

It was long known that mixing two or more monomers together in the same reaction vessel could produce a copolymer. Copolymers offer the opportunity to design a specific set of properties that cannot be obtained from any single homopolymer. For example, polystyrene is a weak, brittle plastic with poor solvent resistance and poor heat resistance. Copolymerizing styrene with acrylonitrile strengthens the polymer and improves resistance to solvents. Styrene-acrylonitrile copolymers (SAN) are sold in very large volumes and are used in furniture and appliances. Copolymerizing styrene with butadiene imparts rubbery properties, as we have seen. Buna-S and GR-S are good examples. Combining all three of these monomers into terpolymers produces materials that are tough, resistant to solvents, and able to withstand relatively high temperatures. They are used in applications as diverse as drain pipes, sporting goods, luggage, automobiles, appliances, and toys. We should point out that acrylonitrile-butadiene-styrene (ABS) materials are usually not linear and random terpolymers, but are prepared with complex structures that optimize the properties of each monomer. In other words, the ultimate properties of a polymer are dependent upon the monomer(s) from which it is made as well as the techniques used in its preparation.

The period roughly from the Second World War to the 1960s was noted for the introduction of a number of commercially important copolymers. Early pioneers such as Hermann Staudinger and Paul Flory recognized that copolymerization reactions could be somewhat complicated. For one thing, the composition of a copolymer was not always the same as that of the starting monomer mixture. Scientists have developed mathematical tools that model copolymerization reactions, aiding the development of a significant number of commercially important copolymers.

Instead of polymerizing two or more monomers together to make co-polymers, could we simply mix homopolymers together and achieve the same thing? The answer is "yes and no." Yes, we can mix polymers together and make *polymer blends*. No, the properties would be somewhat different or, in many cases, simply awful. We will encounter blends again in Chapter 5.

Toward New Architectures

Polymerization of monomers using metal catalysts began as early as 1925, when German scientists began polymerizing butadiene in the presence of metallic sodium, attempting to prepare synthetic rubber. As we have seen, over the next three decades, Ziegler, Natta, and many other scientists studied the reaction of monomers with alkali metals and metal-organic compounds. In 1956, Michael Szwarc and coworkers at Syracuse University discovered a new reaction that became known as "living" polymerization. In this process, which we will discuss in Chapter 5, the reactive intermediate by which the polymer chain "grows" has an unusually long lifetime. Because of this, Szwarc was able to use this new chemistry to synthesize *block copolymers*. He poly-merized the first block, say a polystyrene block, that would stop growing when the styrene monomer had all reacted, but which still had a reactive intermedi-ate at the end of each block. On adding a second monomer, say isoprene, the reaction would continue, but now a block of polyisoprene would grow on each polystyrene chain. The resulting product [poly(styrene-block-isoprene)] has significantly different properties from those of a random copolymer (such as Buna-S rubber, discussed above). This event was significant, because it was basically the start of a period in which the *chain architecture* of polymers could be systematically controlled. Many of the unique properties of modern polymers can be traced to their molecular architecture. Thus began an inter-esting and important chapter in the development of polymer science and engi-neering that continues to this day.

Conducting Polymers

It was soon realized that metal-organic catalysts could be used to prepare other types of polymers, including polymers that can conduct electricity. Most polymers, just like most organic compounds, are electrically insulating. Electrical conduction requires the movement of either electrons or ions. The electrons in saturated organic compounds (no double or triple bonds) are paired and reside in stable, relatively unreactive orbitals. Recall that polyethylene's very high insulative property was the key to the development of practical radar during World War II. If we blend polyethylene with something that conducts electricity, such as a metal, graphite, or carbon fiber, the composite material will be a conductor, and these kinds of composites find many useful applications. In this case the polymer simply provides the matrix for the conductor but itself remains an insulator.

What would it take to make the polymer itself conducting? To start, we would need to make a polymer with a long series of interacting double bonds (called *conjugated unsaturation*). Three examples are polyacetylene, poly(p-phenylene), and polythiophene:

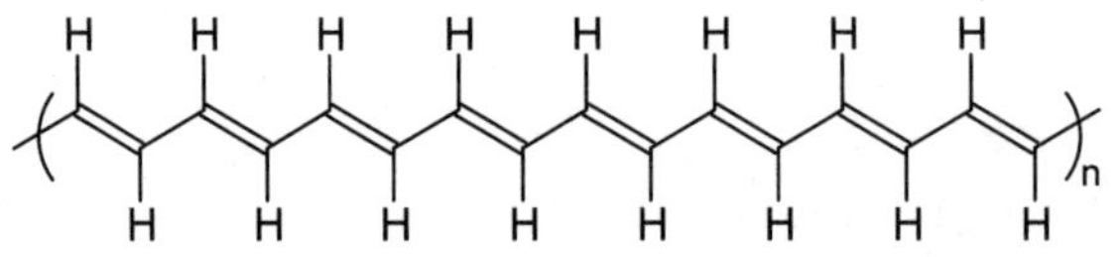

polyacetylene

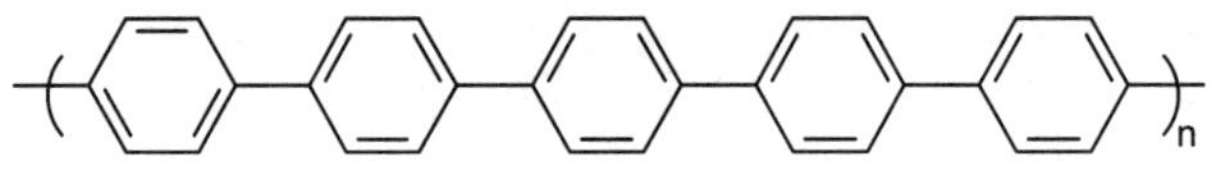

poly(p-phenylene)

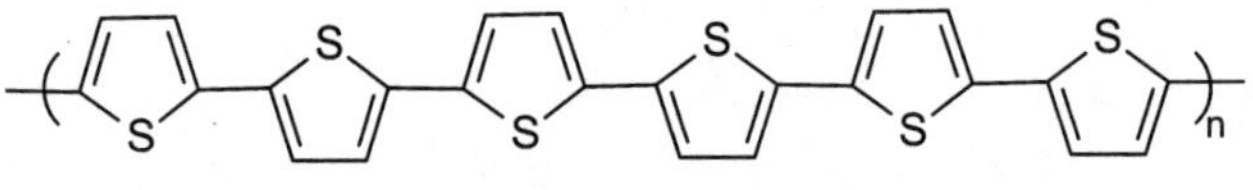

polythiophene

Figure 4-2 shows the huge range of conductivities for various materials, including some polymers. (*Conductivity*, the ability of a material to conduct electrical charge, has the units siemans/cm. A sieman is a reciprocal ohm.) We can divide the series into insulators, semiconductors, and conductors. Polymers such as the three above are not inherently conducting by themselves, but become so when a chemical *dopant* such as iodine, AsF_5, or even an acid is added. Because the doped polymers can be processed into a variety of shapes and have low mass, they are used in small batteries, light-emitting diodes, thin-film transistors, and in antistatic applications. The development of the field of electrically conducting polymers, spearheaded by Alan MacDiarmid,

72

Figure 4-2. Range of electrical conductivities in Siemans/cm for various inorganic and poly-meric materials.

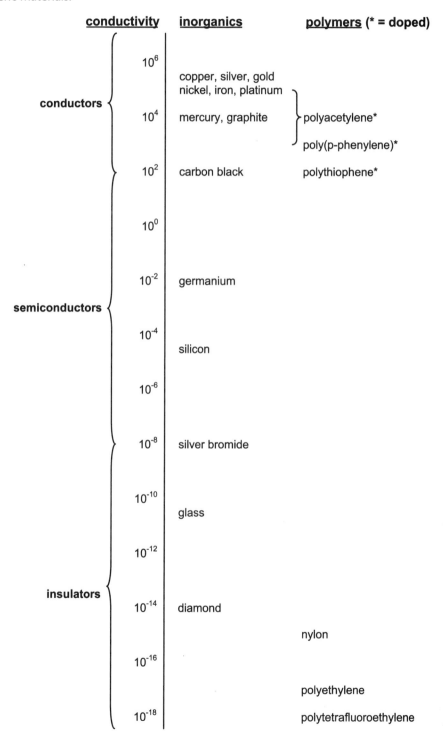

Alan Heeger, and Hideki Shirakawa in the mid 1970s, resulted in their sharing the Nobel Prize in chemistry in 2000.

The Need for Polymers in Electronics

As we mentioned at the start of this chapter, without synthetic polymers the electronics industry as we know it would not exist. Let's find out why. Polymers that can withstand physical abuse and elevated temperatures have been used for many years as the base for circuit boards or as protective coatings for electronic components. Common examples include epoxy resins and polyimides. These applications, while important, are pretty mundane, really. With the development of solid-state devices and integrated circuits (IC), a need for their easy fabrication arose. For this, polymers were needed whose properties changed upon exposure to light (*photopolymers*). In preparing a solid-state electronic device on a silicon chip, three-dimensional circuit elements (conductors, transistors, diodes, etc.) are fabricated in a repetitive series of steps called microlithography or photoengraving, adapted from an old standard printing process (Turner and Daly 1987).

Figure 4-3 (not drawn to scale) illustrates a typical process in which tiny areas of silicon are doped to produce the desired electrical conductivity. First, a photopolymer called a *photoresist* is coated onto a silicon wafer that contains a thin layer of silicon dioxide on it. Ultraviolet light (or other electromagnetic radiation) passes through a *mask* or stencil that contains the electronic patterns and is focussed onto the photoresist layer, causing the polymer that was exposed to undergo a chemical change. Two basic types of photoresists exist. In the first, called a *positive-working resist*, the irradiated polymer becomes more soluble, making it easy to be removed in a subsequent washing step. The increase in solu-

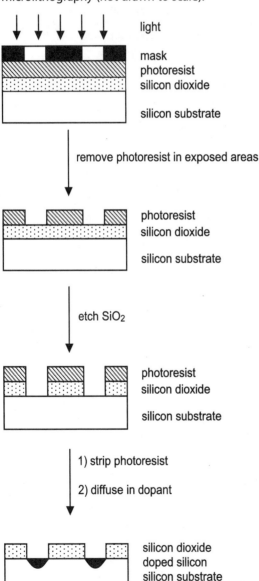

Figure 4-3. Positive-working photoresist for microlithography (not drawn to scale).

light

mask
photoresist
silicon dioxide

silicon substrate

remove photoresist in exposed areas

photoresist
silicon dioxide

silicon substrate

etch SiO$_2$

photoresist
silicon dioxide

silicon substrate

1) strip photoresist

2) diffuse in dopant

silicon dioxide
doped silicon
silicon substrate

bility of the photoresist might be caused by a large decrease in molar mass during exposure (degradation). Alternatively, it might be caused by the loss of a functional group on exposure, producing a change in solubility. Following exposure, the wafer is *developed* with a solvent system that selectively dissolves and removes the polymer only from the exposed areas, revealing the SiO_2 underneath. This is called a positive image because channels are created where light passed through the mask. Figure 4-3 depicts the basic image transfer steps using a positive-working photoresist.

The other type of photoresist is called *negative-working*. Exposure of a negative-working resist causes a decrease in solubility, usually the result of a chemical crosslinking. As we will see in Chapter 6, chemically crosslinked polymers cannot dissolve in any solvent. Development of this system removes polymer from unexposed areas, producing a negative image on the wafer.

Both processes create channels that are then chemically *etched* to remove the SiO_2 and expose the bare silicon underneath. Only the exposed SiO_2 is removed, the photoresist polymer preventing its removal elsewhere (thus the term *resist*). Finally, the wafer is subjected to a *doping agent* that adds a tiny impurity in the exposed areas to change the conductivity of the silicon in these areas. The entire process is repeated a number of times to build up three-dimensional devices.

Since the first chips became commercially available more than three decades ago, the engineers and polymer chemists have continually advanced the technology so that the devices have become smaller and smaller and much more intricate, enabling more devices per chip. The maximum number of transistors on a single microprocessor chip has gone from approximately 2000 in 1971 to more than 30 million today! One of the primary factors that has made this possible is a dramatic drop in the minimum dimension of a feature (e.g., a groove)—from greater than 20 microns in 1963, to less than 2 microns in 1983, to less than 0.2 microns (200 nm) in 2003. Devices are much closer together, meaning processes require much less time. If the speed of automobiles had increased in a similar way, it would take only about 13 minutes to drive from New York City to San Francisco (Anon-Intel 2002). Of course, the cars would be much smaller, too.

Not all polymers are suitable for use as photoresists. The polymer must initially dissolve in an organic solvent so that it can be coated onto the wafer, and it must form a good film. It must contain a functional group that either responds directly to the irradiating energy or to a light-sensitive additive. Finally, it must be removed (either where it was irradiated or where it was not irradiated) completely, leaving behind a feature with the proper geometry and mechanical properties to protect the layer(s) underneath. The earliest materials were "old" polymers that had been used in the printing industry. Chief among them were novolacs, the phenol-formaldehyde resins invented by Baekeland. The first commercial negative photoresists were introduced in 1954 by Eastman Kodak Company for use in the manufacture of printed circuit boards.

Throughout the text, I have tried to make the point that polymer science is a discipline that crosses many traditional boundaries. We can see this clearly in talking about specialty materials such as photoreactive polymers. Organic and inorganic chemists have long known of compounds that undergo physical and chemical changes in the presence of light. (There is probably no more sophisticated example of this than chlorophyll.) The electrical engineer trying to design an integrated circuit device needs a procedure to manufacture it. He or she wants a material that can be coated in thin layers from solvent, that has physical integrity, that undergoes a change in the presence of light, and that can be removed with another solvent. The polymer chemist designs a polymer that combines these different features. He or she will synthesize a number of examples, consult with the engineer and the people fabricating the device, and work through the inevitable problems to optimize a composition.

If this is a project sponsored by a for-profit company, the polymer chemist will need a synthetic procedure that is practical, safe, and environmentally benign, that can be scaled up for manufacturing, and that is inexpensive enough so the company can make a profit. It should come as no surprise that as the electronics industry continues to produce smaller and more sophisticated devices, the demand for new ideas and new materials remains extremely high. For one, a plastic semiconductor stable in air (Anon. CEN 2002) or a polymeric transistor (Dagani 2001) that could replace silicon would eliminate the need for complicated and relatively expensive silicon fabrication technology.

The Influence of Sputnik

On October 5, 1957, the same year of the Hula Hoop craze, much of the world was shocked at the announcement that the Soviet Union had successfully launched into space Sputnik 1, the world's first satellite to orbit the Earth. By that time, the United States and the Union of Soviet Socialist Republics, the then-USSR, had been immersed for several years in a tense Cold War, with the threat of nuclear attack hanging over everyone's head. Thus, not only did Sputnik raise real military fears, but it also suggested that Soviet science and technology were superior to those in the United States, initiating a significant number of governmental and educational changes. It changed the way chemistry, physics, and mathematics were taught. And it generated a large program in aerospace research and technology that resulted in Apollo 11 landing the first human on the Moon 12 years later.

The extreme conditions of space demanded high-performance materials that could withstand them. And lifting large payloads off the launch pad would require conveyance vehicles constructed of lightweight materials.

Demands such as these brought about a new materials research effort that eventually produced whole new classes of polymers and much greater understanding of polymer science. New lightweight fibers such as carbon fibers were developed, as well as the extremely strong polyaramids (e.g., Kevlar). More recently, many of the structural members of the B-2 stealth bomber are fabricated from composites of high-performance polymers and carbon fiber rather than from the more traditional aluminum or titanium.

One of the pioneers in designing high-temperature, high-performance polymers was Carl "Speed" Marvel (1894–1988). Marvel obtained his PhD in organic chemistry from the University of Illinois, where he then spent 41 years as a faculty member in the chemistry department. One of the first students in his organic class was Wallace Carothers, who was just beginning his doctoral studies. Marvel soon became interested in polymers, and was probably the first person to synthesize linear polyethylene, although he did not pursue it. This was 1928, and no one thought that polyethylene would be at all useful. During World War II Marvel was a key member of the GR-S program, where he contributed both practical ideas and fundamental understanding. In the 1950s and 1960s Marvel turned his attention to polymers that could withstand very high temperatures, materials that could be useful for supersonic flight and space exploration. He and his students synthesized step-growth polymers with rigid backbones made up of thermally stable rings. This was a new class of materials that are tough, have high melting points, have the ability to resist attack by solvents, and do not readily burn. These discoveries led to the development of several new classes of polymers for high-performance applications in the following decades. Forced to retire from the University of Illinois at age 70 in 1961, Marvel continued his productive career at the University of Arizona for another two decades (Marvel 1987).

The period of molecular engineering had begun. The high-temperature, high-performance polymers generally exceeded the properties of what are known as engineering polymers and tended to be very expensive. Over time, however, the cost of many of them has decreased so that they are now chosen for more mundane applications such as commercial aircraft and automobile engines. Some of the materials are now commonly used in sporting equipment (e.g., golf club shafts, pole vaulting poles, skis) and in civil construction. Figure 4-4 shows the longest "plastic" bridge in the United States when it was built in 2001. It spans an outlet of one of the Finger Lakes in Upstate New York. Although it weighs approximately one-quarter of a traditional steel and concrete bridge, it has no posted weight limit and should require only minimal maintenance for at least 75 years. It is not affected by the weather, nor is it corroded by road salt. Traditional bridges, especially in the North, need periodic painting and usually require significant repair within 20 years (Spector 2001).

Figure 4-4. Installation of "longest plastic bridge in the United States."

Summing Up

The history of polymer science begins with the attempt to modify natural polymers to suit human needs, to improve upon nature. During much of the nineteenth century, significant, often fortuitous discoveries resulted in technological breakthroughs and a host of life-altering products—rubber, new textile fibers, amateur photography, and motion pictures. Celluloid, the first plastic, was produced in this period, highly desired because it replaced the increasingly rare ivory. As a new century was just beginning, the first polymer synthesized solely from small molecules was invented. Bakelite became the "material of a thousand uses."

By the early twentieth century, scientists were arguing about the very existence of macromolecules, a completely radical idea at the time. The continued synthesis of polymers, often fueled by the demands of war, brought even better materials and newer technologies. The fascination with polymers by Staudinger and Mark eventually drew the attention of others, and an interdisciplinary science was born. Staudinger the organic chemist recognized that it was the shape of the macromolecules themselves that dictated the physical and chemical properties of polymers. Carothers, too, brought a molecular bias to his work, realizing that very long chains of organic repeat units would have unusual and useful properties. All of these scientists drew heavily on their understanding of natural polymers.

The Second World War, and the years following it, saw increasing efforts to produce ever more sophisticated materials. Organic chemists, physical chemists, physicists, and engineers were spending their careers advancing the field of polymer science while new institutes devoted to the formal education of polymer scientists formed. Better understanding of the chemistry of polymer formation sparked the discovery of alternative synthetic methods and, in turn, even better materials. Ziegler-Natta stereoregular polymerization began a cascade of olefin polymerizations that continues to this day. Engineering polymers, stronger and more heat-resistant materials, found applications in automobiles, airplanes, and appliances. Space exploration placed much higher demands for new, lightweight materials, prompting efforts to produce high-performance materials. Polymers with magnetic properties or which conduct electricity or light energy continue to become more sophisticated and to find new applications. And, in just the same way as we started, we are still trying to improve on nature. The integration of traditional (synthetic) polymer science with biological fields is a rapidly growing area, one that will change our lives in new and unexpected ways in the future.

References Cited

Anon. 2001. Manmade fiber production tops 31 million tons as natural fibers post slight decline. *International Fiber Journal* 16 (4) (August).

Anon. CEN. 2002. Xerox claims electronics polymer advance. *Chemical and Engineering News* December 9: 11.

Anon-Intel. 2002. *http://www.intel.com/intel/intelis/museum/exhibits/index.htm.*

Dagani, R. 2001. Polymer transistors: Do it by printing. *Chemical and Engineering News* January 1: 26–27.

Fenichell, S. 1996. *Plastic: The making of a synthetic century.* New York, NY: HarperBusiness.

Flory, P. J. 1953. *Principles of polymer chemistry.* Ithaca, NY: Cornell University Press.

Guillet, J. 1974. Plastics, energy, and ecology—a harmonious triad. *Plastics Engineering* August: 48–56.

Kauffman, G. B. 1993. Rayon: The first semi-synthetic fiber product. *Journal of Chemical Education* 70 (11): 887–93.

Kiefer, D. M. 2002. Review of *The madman who made rubber useful,* by R. Korman. *Chemical and Engineering News* August 12: 44–45.

Marvel, C. S. 1987. The development of polymer chemistry in America – the early days. *Journal of Chemical Education* 58(7): 535–39.

McMillan, F. M. 1979. *The chain straighteners. Fruitful innovation: The discovery of linear and stereoregular synthetic polymers.* London: Macmillan Press.

Morawetz, M. 1985. *Polymers: The origins and growth of a science.* New York, NY: Wiley Interscience.

Morris, P. J. T. 1986. *Polymer pioneers: A popular history of the science and technology of large molecules.* Philadelphia, PA: Center for History of Chemistry Polymer Project.

Raber, L. R. 2001. Landmark honors Carothers' work. *Chemical and Engineering News* January 22: 108–09.

Spector, J. 2001. Plastic bridge put in its place. Rochester, NY *Democrat and Chronicle* December 13: 1B.

Staudinger, H. 1970. *From organic chemistry to macromolecules: A scientific autobiography based on my original papers.* New York, NY: Wiley Interscience.

Turner, S. R., and R. C. Daly. 1987. Polymers in microlithography. *Journal of Chemical Education* 65(4): 322–25.

Other Reading

Alvino, W. M. 1995. *Plastics for electronics: Materials, properties, and design applications.* New York, NY: McGraw-Hill Book Company.

Brayer, E. 1996. *George Eastman: A biography.* Baltimore, MD: Johns Hopkins University Press.

Emsley, J. 1998. *Molecules at an exhibition: Portraits of intriguing materials in everyday life.* Gallery 5. New York, NY: Oxford University Press.

Hermes, M. E. 1996. *Enough for one lifetime: Wallace Carothers, inventor of nylon.* Washington, DC: American Chemical Society and the Chemical Heritage Foundation.

Hochheiser, S. 1988. The development of Plexiglas in the United States. *Today's Chemist* June: 8–10.

Hounshell, D. A., and J. K. Smith, Jr. 1988. The nylon drama. *Invention and Technology* Fall: 40–55.

Mark, H. 1973. The early days of polymer science. *Journal of Chemical Education* 50 (11): 757–60.

Mark, H. 1987. From revolution to evolution. *Journal of Chemical Education* 64 (10): 858–61.

Nobel Prizes in chemistry. *www.nobel.se/chemistry/laureates*

Seymour, R. B., and C. E. Carraher. 1990. *Giant Molecules: Essential materials for everyday living and problem solving.* New York: Wiley Interscience.

Seymour, R. B., and G. B. Kauffman. 1992. The rise and fall of celluloid. *Journal of Chemical Education* 69 (4): 311–14.

Sherman, B. C., W. B. Euler, and R. R. Forcé. 1994. Polyaniline—A conducting polymer; electrochemical synthesis and electrochromic properties. *Journal of Chemical Education* 71 (4): A94–A95.

Winfield, A. G. 1992. Fifty years of plastics. *Plastics Engineering* May: 32–47.

Section 2 Synthesis and Properties

Polymer Synthesis

> *Without synthetic polymers, (there is) no standard of living.*
> Dr. Hans Uwe Schenck

I n the first four chapters, we have seen that many different kinds of polymers exist and that they have an extremely wide range of properties. Some are stiff, others are soluble, while still others are rubbery. There are plastics, and fibers, and adhesives, and foams. The structure and composition of the macromolecule dictate the ultimate properties. Structure and composition are determined when the macromolecule is synthesized.

How Do They Grow?

In this chapter we want to better understand how monomers react together to form long polymer chains. We have already seen a few reactions of organic compounds. For example, in Chapter 4 we wrote an equation for the esterification reaction of an alcohol with a carboxylic acid to produce an ester plus water (Equation 5). We pointed out that monomers are usually difunctional organic compounds, where reaction with other suitable difunctional compounds can lead to polymer formation. In Chapter 4 we illustrated this with a polyesterification reaction (Equation 6). We will see that there are several different types of monomers. After reading this chapter, you should be able to identify those organic compounds that are monomers and understand how they can react to form polymers.

In Chapter 2, we identified three different types of compounds that are commonly converted to polymers:

- compounds with carbon-carbon double bonds
- certain cyclic compounds
- compounds with two or more functional groups that react to form a new functional group, often with the evolution of some small molecule such as water, methanol, or HCl

Examples of each type were given in Tables 2-2 and 2-3. It is now time to better understand *how* compounds such as these react to form macromolecules.

Some 50 years ago, Paul Flory chose the terms *step-growth* and *chain-growth* polymerization to describe the processes by which many monomers are converted to polymer (Flory 1953). Although not perfect, the terms are still commonly used and can help us understand the major mechanisms of polymerizations. A *mechanism* for a reaction describes the processes and pathways by which that reaction proceeds. Mechanisms are important because they help us understand the details of a chemical reaction as well as help us predict the outcome of new reactions.

A Comparison of Step-Growth and Chain-Growth Polymerizations
Step-Growth

Step-growth polymerization begins with the stepwise reaction of two difunctional monomers, forming an intermediate compound with a new functional group. The intermediate then reacts with another monomer molecule, or with another intermediate molecule, forming a larger intermediate. These steps continue until high polymer is obtained. Many of these reactions also produce a small-molecule byproduct (i.e., condensation reactions), in which case they are called polycondensations. This is illustrated below for the polymerization of the methyl ester of lactic acid, methyl lactate, in which the hydroxy group on the monomer reacts with the ester group on another molecule, forming a new ester and releasing the byproduct methanol:

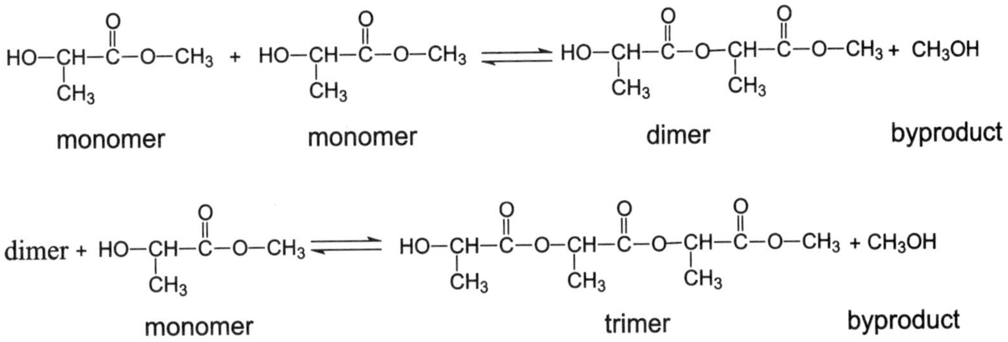

$$HO-\underset{\underset{CH_3}{|}}{CH}-\underset{\underset{}{\overset{\overset{O}{\|}}{C}}}-O-CH_3 \;+\; HO-\underset{\underset{CH_3}{|}}{CH}-\underset{}{\overset{\overset{O}{\|}}{C}}-O-CH_3 \;\rightleftharpoons\; HO-\underset{\underset{CH_3}{|}}{CH}-\overset{\overset{O}{\|}}{C}-O-\underset{\underset{CH_3}{|}}{CH}-\overset{\overset{O}{\|}}{C}-O-CH_3 \;+\; CH_3OH$$

      monomer   monomer   dimer   byproduct

$$dimer + HO-\underset{\underset{CH_3}{|}}{CH}-\overset{\overset{O}{\|}}{C}-O-CH_3 \;\rightleftharpoons\; HO-\underset{\underset{CH_3}{|}}{CH}-\overset{\overset{O}{\|}}{C}-O-\underset{\underset{CH_3}{|}}{CH}-\overset{\overset{O}{\|}}{C}-O-\underset{\underset{CH_3}{|}}{CH}-\overset{\overset{O}{\|}}{C}-O-CH_3 \;+\; CH_3OH$$

           monomer   trimer   byproduct

trimer + monomer $\rightleftharpoons$ tetramer + byproduct

This particular reaction is called *ester exchange*, or *transesterification*—the conversion of one ester into another. Because it is a reversible reaction, the methanol byproduct needs to be removed from the reaction mixture as the polymerization proceeds in order to drive the reaction toward high polymer.

Note that the intermediates formed during each step of the polymerization are similar to the monomer—they contain an OH group at one end and a methyl ester group at the other. In addition, of course, the intermediates contain other ester groups, because we are making a *poly*ester. *All* of these functional groups are susceptible to reaction. The scheme below includes the kinds of reactions that would lead eventually to polymer:

dimer	+	dimer	$\longrightarrow$	tetramer
trimer	+	dimer	$\longrightarrow$	pentamer
trimer	+	trimer	$\longrightarrow$	hexamer
pentamer	+	monomer	$\longrightarrow$	hexamer
tetramer	+	tetramer	$\longrightarrow$	octamer
50-mer	+	37-mer	$\longrightarrow$	87-mer
etc.			$\longrightarrow$	polymer

During the course of the polymerization, monomer disappears relatively quickly, the average molar mass of the intermediate molecules increases gradually, and high molar mass polymer is formed only at the very end. The average degree of polymerization and the average molar mass increase slowly as the reaction proceeds. The final molar mass is determined primarily by how far the polymerization reaction proceeds. This is illustrated in Figure 5-1.

Figure 5-1. Polymer molar mass as polymerization proceeds by step-growth mechanism.

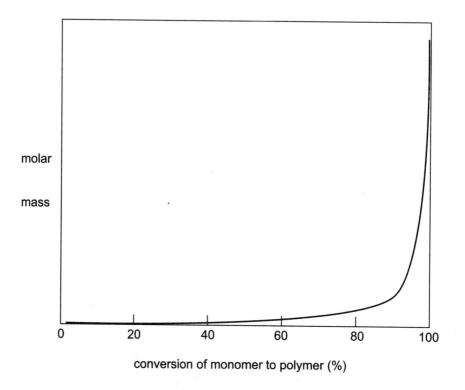

Many of our common polymers are made by this type of reaction, including polyesters, nylons, polycarbonates, polyurethanes, and polyaramids (e.g., Kevlar). Many natural polymers, including cellulose, starch, proteins, and polynucleotides are also step-growth polymers.

Chain-Growth

We have said several times that monomers are usually difunctional molecules. Molecules that undergo step-growth polymerization are clearly difunctional—they contain two functional groups, either the same or different. When we think of chain-growth polymerization, however, the first compounds that come to mind are those containing one carbon-carbon double bond, such as ethylene, propylene, or styrene. Sometimes monomers with a carbon-carbon double bond are loosely called *vinyl* monomers. What makes a compound containing only one double bond difunctional? How can we react molecules of something like ethylene together and make new carbon-carbon single bonds?

For one thing, we know that compounds containing a carbon-carbon double bond (*unsaturated* compounds) can undergo addition reactions and become *saturated*. For example, one mole of a compound containing one carbon-carbon double bond will add one mole of compounds such as hydrogen, hydrogen chloride, bromine, or water. In Equation 4, the species adding to the double bond is a molecule of bromine consisting of two bromine atoms.

Equation 4

$$CH_2\!\!=\!\!CH_2 \quad + \quad Br\!\!-\!\!Br \quad \longrightarrow \quad Br\!\!-\!\!CH_2\!\!-\!\!CH_2\!\!-\!\!Br$$

We can think of a double bond as being made up of a sigma bond and a pi bond, each containing two electrons and each being shared by the two carbon atoms making up the double bond. The pi electrons are reactive, and can reorganize to form two new sigma bonds with *two* other species. The two carbon atoms change from sp^2 hybridization in ethylene to sp^3 in the saturated product, 1,2-dibromoethane.

However, ethylene molecules do not react directly with each other to form polyethylene. We must first convert a molecule of ethylene to a *reactive intermediate*, a chemically reactive species that can react with a second molecule of ethylene, forming a new, four-carbon reactive intermediate. This is shown in Equations 5 and 6. In Equation 5, some initiating, reactive species attacks a molecule of ethylene, producing the new reactive intermediate. In equation 6, this intermediate attacks a second molecule of ethylene, producing a new carbon-carbon bond and generating a larger reactive intermediate. Monomer can add only to a reactive intermediate, not to another monomer. This chain reaction continues until some reaction occurs that breaks the chain.

Equation 5

$$R^* \quad + \quad CH_2\!\!=\!\!CH_2 \quad \longrightarrow \quad R\!\!-\!\!CH_2\!\!-\!\!CH_2^*$$

Equation 6

$$R\!\!-\!\!CH_2\!\!-\!\!CH_2^* \quad + \quad CH_2\!\!=\!\!CH_2 \quad \longrightarrow \quad R\!\!-\!\!CH_2\!\!-\!\!CH_2\!\!-\!\!CH_2\!\!-\!\!CH_2^*$$

Reactions such as this are significant because new carbon-carbon bonds are formed. It is easy to manipulate various functional groups: acids to esters or amides, alcohols to ethers or aldehydes or ketones, etc. But reactions that produce new carbon-carbon bonds in specific ways fall into a special category. We will present the details of chain-growth polymerization shortly. But before we do, let's compare some of the features of this mechanism with those of step-growth polymerization.

Comparison of the Two Reactions

How different these two types of polymerizations are. In step-growth polymerization, monomers with different functional groups react with each other to form larger molecules. Although catalysts are often used to increase the rates of reaction, no initiators are needed to begin the chemistry. In chain-growth polymerization, on the other hand, a monomer first reacts with an initiator to form a reactive species. The reactive species or reactive intermediate (e.g., a cation, a free radical, or an anion) reacts with a molecule of monomer forming another reactive intermediate, which reacts with another monomer, and so on in a *chain reaction*. The reaction continues until, as we said, some side reaction destroys the reactive intermediate and breaks the chain. Unlike step-growth, high polymer is formed early in the reaction. The intermediates are not molecules but rather reactive, short-lived species that react rapidly with monomer. The reaction mixture consists of monomer and

Figure 5-2. Polymer molar mass as polymerization proceeds by chain-growth versus step-growth mechanisms.

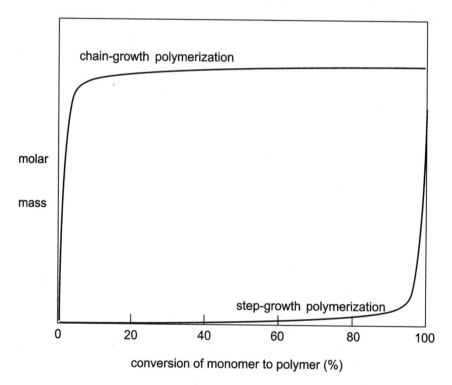

polymer, in addition to very low concentrations of initiator. Plots of polymer molar mass versus conversion of monomer to polymer for the two mechanisms are shown in Figure 5-2. The major differences between these two basic polymerization mechanisms are summarized in Table 5-1.

Table 5-1. Comparison of two major polymerization types.

Step-Growth Polymerization	Chain-Growth Polymerization
monomers have two reactive functional groups	most common monomers have a carbon-carbon double bond
polymer backbone contains carbon and usually oxygen or nitrogen atoms	polymer backbone usually contains only carbon atoms
functional groups react without initiator; reaction may be catalyzed; reaction often gives off small - molecule byproduct	reaction is usually begun with an initiator
intermediates are compounds (e.g., oligomers) with same two reactive functional groups as monomer	polymerization proceeds by chain reaction through short-lived reactive intermediate (free radical, cation, anion); monomer reacts only with reactive intermediate, not with other monomer
polymer develops at slow rate; high polymer is formed only at the very end	high polymer is formed very rapidly
reaction mixture contains monomer, oligomer, and polymer	reaction mixture consists only of monomer and high polymer
DP and molar mass (MM) increase as reaction proceeds	MM of polymer formed early in the reaction is approximately the same as that formed later on; increasing time increases yield, not MM

The polymerization of some monomers does not fall neatly into either of the mechanisms discussed above. We will take up a few of them (e.g., anionic and coordination polymerizations) after we further develop step-growth and chain-growth polymerizations. Some polymerizations can proceed by either mechanism, depending upon the specific monomer or the reaction conditions. The most notable examples, ring-opening polymerization and some of the newer chemistries, are presented as separate categories toward the end of the chapter.

Step-Growth Polymerization in More Detail

As you might already have noticed, monomers for step-growth polymerization fall into approximately three categories. In many cases, two monomers with different functional groups react together to form a polymer. One of them might be a diacid or a derivative such as a diester. The other monomer might be a dihydroxy compound (a diol) or a diamine. Common polymer examples that result from the polymerization of these kinds of monomers include PET (poly[ethylene terephthalate]) and nylon-6,6 (see Table 2-3). These polymers are homopolymers, not copolymers, even though they are prepared from two different monomers. They are homopolymers because the polymer has only one repeat unit.

The second type of monomer for step-growth polymerization contains two *different* functional groups. Examples in this category include hydroxy acids such as lactic acid (or hydroxy esters [Equations 1–3]), and amino acids. A third type includes cyclic monomers such as lactones, lactams, and cyclic ethers. Cyclic monomers polymerize by ring-opening polymerization. Some, as we said, proceed by step-growth and some by chain-growth mechanisms.

Making PET in the Melt

Polycondensations, as these reactions are sometimes called, are usually run in the absence of solvent. For example, consider the synthesis of PET. Dimethyl terephthalate (DMT) and an excess of ethylene glycol are combined in a stainless steel reactor, a transition metal catalyst is added (a few parts per million), and the mixture heated until it melts. Heating and stirring are continued for a few hours, during which time the ethylene glycol reacts with the DMT, forming bis(hydroxyethyl)terephthalate (transesterification stage) (see Equation 7). During this stage, byproduct methanol is distilled from the reactor. Next, the pressure in the reactor is reduced, the temperature is raised, and the excess ethylene glycol is distilled (polycondensation stage) (see Equation 8). This reaction proceeds until the molar mass of the polymer is high enough (>99% conversion). Then the reactor is returned to atmospheric pressure and the molten PET is pumped out of the reactor, ready for melt extrusion or fiber formation.

A polymerization such as this is called a melt condensation because the reaction is run "neat," meaning no solvents are used. Thus there is no need to purify the polymer at the end, and all leftover monomers are either converted to polymer or distilled from the reactor. Reactions like this are sometimes called *green* because of their relatively low demands on the environment. The yield is 100% and there are no side reaction products that need to be removed. Because no solvents are used, none escape into the air or the sewer. The product does not have to be isolated from a solution, a process that would require energy. The methanol and excess ethylene glycol that are distilled from the

Equation 7

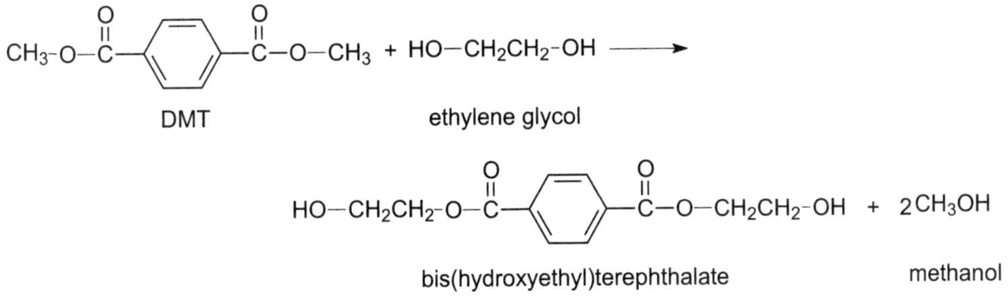

DMT   ethylene glycol

$$HO-CH_2CH_2-O-\overset{O}{\underset{\|}{C}}-\langle\text{benzene}\rangle-\overset{O}{\underset{\|}{C}}-O-CH_2CH_2-OH \ + \ 2CH_3OH$$

bis(hydroxyethyl)terephthalate   methanol

Equation 8

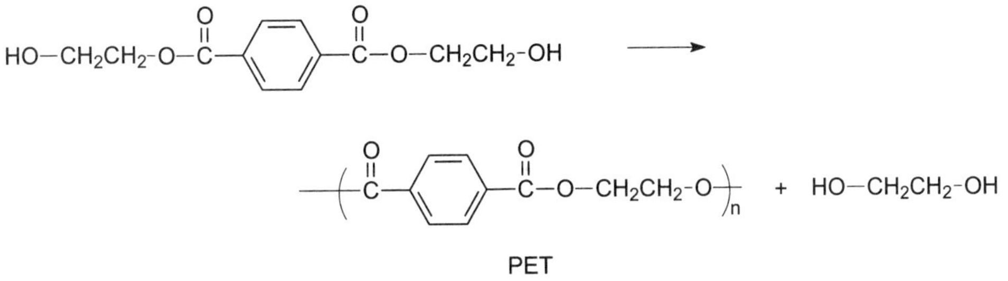

$$HO-CH_2CH_2-O-\overset{O}{\underset{\|}{C}}-\langle\text{benzene}\rangle-\overset{O}{\underset{\|}{C}}-O-CH_2CH_2-OH \longrightarrow$$

$$-\left(\overset{O}{\underset{\|}{C}}-\langle\text{benzene}\rangle-\overset{O}{\underset{\|}{C}}-O-CH_2CH_2-O\right)_n \ + \ HO-CH_2CH_2-OH$$

PET

reactor are condensed and reused. Temperatures required to run melt poly-condensations are often as high as 250° to 275°C. Clearly this is a technique that is suitable only for those polymers (and monomers) that are stable at such high temperatures.

Interfacial Polycondensation

If one were to choose more reactive monomers, it would be possible to carry out polycondensations at considerably lower temperatures in solution. For example, consider the reaction of a diamine and a diacid to make a polyam-ide (nylon), a polymerization that requires relatively high temperatures (see Equation 9). A much faster reaction would occur between the diamine and a corresponding diacid chloride (see Equation 10). Both reactions would pro-duce the same polymer, although the reaction conditions would be much different, and the byproduct HCl from the acid chloride reaction would have to be carefully trapped. One technique for performing a polymerization such as that in Equation 10 is to dissolve the monomers in different, immiscible solvents, forcing the polymerization to occur only at the interface of the two solvents, a process called *interfacial polymerization*. Because of the high re-activity of an acid chloride, these reactions can be carried out at very low temperatures. This polymerization can be carried out rather dramatically in a beaker and is known as the "nylon rope trick" (see Section 4).

Equation 9

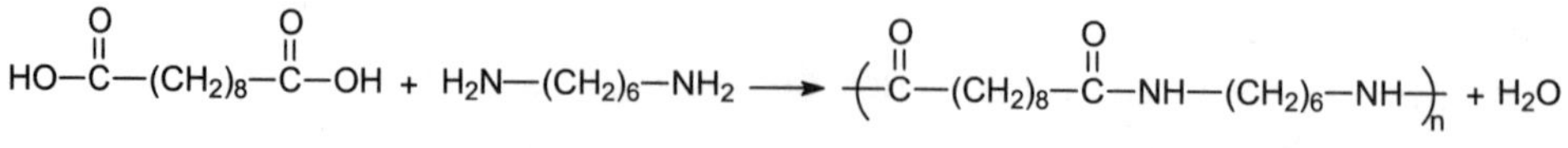

$$HO-\overset{O}{\underset{||}{C}}-(CH_2)_8-\overset{O}{\underset{||}{C}}-OH \ + \ H_2N-(CH_2)_6-NH_2 \ \longrightarrow \ \left(\overset{O}{\underset{||}{C}}-(CH_2)_8-\overset{O}{\underset{||}{C}}-NH-(CH_2)_6-NH\right)_n \ + \ H_2O$$

Equation 10

$$Cl-\overset{O}{\underset{||}{C}}-(CH_2)_8-\overset{O}{\underset{||}{C}}-Cl \ + \ H_2N-(CH_2)_6-NH_2 \ \longrightarrow \ \left(\overset{O}{\underset{||}{C}}-(CH_2)_8-\overset{O}{\underset{||}{C}}-NH-(CH_2)_6-NH\right)_n \ + \ HCl$$

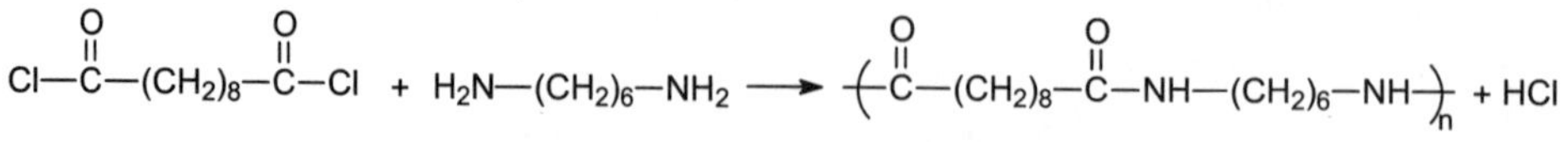

Interfacial polymerization can be used to make many types of step-growth polymers such as polyamides, polyesters, polycarbonates, and polyurethanes. Although most step-growth polymers are prepared by a melt process, some specialty polymers are prepared by the interfacial technique, allowing rapid reaction at low temperatures.

Chain-Growth Polymerization in More Detail
Free Radical Chain Polymerization

Now it's time to learn more about the chemistry of chain-growth reactions. To begin, let's consider the most prevalent type of chain reaction, free radical polymerization. A free radical is defined as a species having one unpaired electron. For example, the methyl free radical would look like this:

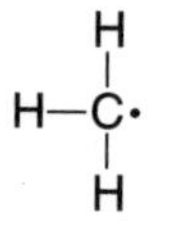

$$H-\overset{H}{\underset{H}{C}}\cdot$$

We take this symbolism to mean that the carbon atom is sharing two electrons with each hydrogen atom and has one unshared electron. This intermediate has no charge, but is very reactive. It can, for example, couple or combine with another radical, pairing both lone electrons and forming a new single bond making the molecule ethane:

Equation 11

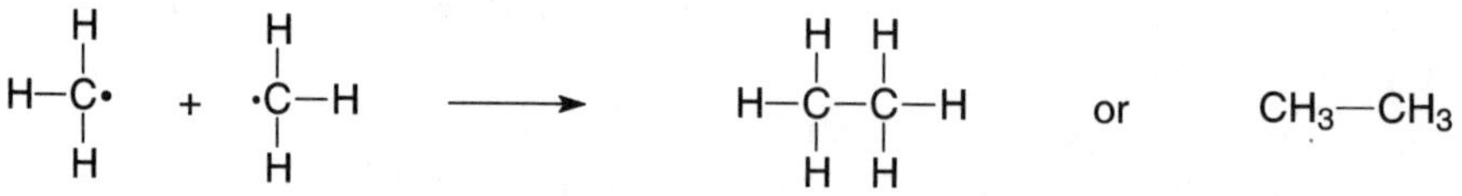

$$H-\overset{H}{\underset{H}{C}}\cdot \ + \ \cdot\overset{H}{\underset{H}{C}}-H \ \longrightarrow \ H-\overset{H}{\underset{H}{C}}-\overset{H}{\underset{H}{C}}-H \quad or \quad CH_3-CH_3$$

A carbon radical can also react with a compound containing a double bond (a monomer), forming a new single bond and generating a new, larger radical. Consider the reaction of a methyl radical with ethylene:

Equation 12

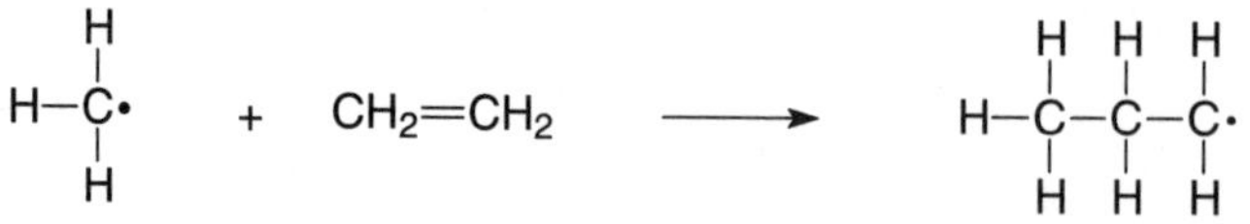

$$H-\overset{H}{\underset{H}{C}}\cdot \ + \ CH_2{=}CH_2 \ \longrightarrow \ H-\overset{H}{\underset{H}{C}}-\overset{H}{\underset{H}{C}}-\overset{H}{\underset{H}{C}}\cdot$$

We can describe this as the lone electron from the methyl radical pairing with one of the electrons of the double bond in ethylene forming the new single bond between the methyl carbon and carbon-1 of ethylene, and the other electron from the double bond becoming unshared on carbon-2.

The mechanism for free radical (or sometimes just radical) chain-growth polymerization involves three steps: radical initiation, chain propagation, and chain termination. We need to have a technique of producing radicals in a controlled manner. Being very reactive, the radicals that form will react with functional groups such as carbon-carbon double bonds, forming new radicals, which will propagate rapidly by reacting with more double bonds. At some point, the growing radical will undergo a reaction that destroys the radical, breaking the chain and terminating its growth. The three major steps for the polymerization of ethylene are shown in the following scheme:

$$I \longrightarrow R\cdot$$

$$R\cdot \;+\; CH_2{=}CH_2 \longrightarrow R{-}CH_2{-}CH_2{\cdot}$$

initiation

$$R{-}CH_2{-}CH_2{\cdot} \;+\; CH_2{=}CH_2 \longrightarrow R{-}CH_2{-}CH_2{-}CH_2{-}CH_2{\cdot}$$

$$R{-}CH_2{-}CH_2{-}CH_2{-}CH_2{\cdot} \;+\; CH_2{=}CH_2 \longrightarrow R{-}(CH_2{-}CH_2)_2{-}CH_2{-}CH_2{\cdot}$$

....

$$R{-}(CH_2{-}CH_2)_m{-}CH_2{-}CH_2{\cdot} \;+\; CH_2{=}CH_2 \longrightarrow R{-}(CH_2{-}CH_2)_{m+1}{-}CH_2{-}CH_2{\cdot}$$

propagation

Coupling

$$R{-}(CH_2{-}CH_2)_p{-}CH_2{-}CH_2{\cdot} \;+\; {\cdot}CH_2{-}CH_2{-}(CH_2{-}CH_2)_r{-}R \longrightarrow$$

$$R{-}(CH_2{-}CH_2)_p{-}CH_2{-}CH_2{-}CH_2{-}CH_2{-}(CH_2{-}CH_2)_r{-}R$$

Disproportionation

$$R{-}(CH_2{-}CH_2)_p{-}CH_2{-}CH_2{\cdot} \;+\; {\cdot}CH_2{-}CH_2{-}(CH_2{-}CH_2)_r{-}R \longrightarrow$$

$$R{-}(CH_2{-}CH_2)_p{-}CH_2{-}CH_3 \;+\; CH_2{=}CH{-}(CH_2{-}CH_2)_r{-}R$$

termination

"I" stands for the initiator, a chemical species that dissociates into radicals under the influence of heat, light, or an oxidation-reduction reaction. The radical generated reacts with ethylene, forming the initiating species. These steps are called *initiation*. Next, the propagation phase begins, in which the radicals react with monomer and the chain grows two carbon atoms at a time.

Finally, some termination step occurs, two of which are shown in the scheme. The most common is *coupling*, in which two radicals combine, leading to one larger macromolecule. Polystyrene radicals typically undergo termination by coupling. Another reaction that is common with some monomers (e.g., methyl methacrylate) is called *disproportionation* in which on the reaction of two radicals, a hydrogen atom transfers from one species to the other.

During radical chain-growth polymerization, initiating radicals are formed gradually over time, some of which react with monomer, initiating a carbon radical that rapidly propagates to high polymer and then terminates. As we said, the reaction mixture consists of monomer and polymer, plus extremely low concentrations of initiator. Radical polymerizations can be carried out in solution or, in some cases, without solvent (in bulk). Temperature control is very important in radical polymerizations, which are highly exothermic. Heat is often added to a reactor at the start to generate initiating radicals. Once the polymerization begins, however, heat might need to be removed by cooling the reactor to prevent unwanted temperature increase. On a large scale, allowing a monomer to undergo a bulk polymerization could easily lead to a runaway reaction, with the buildup of a dangerously large amount of heat, endangering equipment and anyone standing nearby. Because free radicals can be generated whenever oxygen encounters some organic compounds (including monomers), the transport and storage of large quantities of monomers pose serious safety concerns. Therefore, samples of monomers generally contain *inhibitors*, compounds that intercept and destroy radicals so that chain reactions cannot get started. In the laboratory, small-scale solution and bulk polymerizations can be carried out routinely with little hazard.

On an industrial scale, solution polymerizations can be done safely but are generally not ideal. The solvent dilutes the monomer, slowing the propagation reaction but aiding the transfer of heat out of the reactor. A solution is less viscous ("thick") than a bulk polymer mixture, so is easier to stir. However, using a solvent adds cost and complicates the process. Some solvents interfere with the polymerization, causing a decrease in molar mass or an increase in branching or crosslinking. At the end of the polymerization the solvent must be separated from the polymer and recovered.

It would be useful to have liquid present in the polymerization reactor that provided the advantages of a solvent but without any of the disadvantages. Sound unlikely? How about water? One technique, called *suspension polymerization*, involves adding monomer to water in a reactor, agitating the mixture rapidly so that the monomer breaks apart into very small droplets, adding an initiator that is soluble in the monomer, and heating. Each droplet acts as a microbulk polymerization, the water very effectively removes the heat of polymerization, and the resulting polymer spheres are easily separated and filtered. This process, also known as *bead polymeriza-*

94

tion, is used to produce small beads (0.001 to 1 cm in diameter) of pure polymer.

Going One Step Better: Emulsion Polymerization

Suppose we wanted a polymer that was going to be used for making some kind of a coating, an adhesive, or that was going to be added to some mixture and the solids isolated by drying. For environmental reasons, we do not want to use any organic solvents, meaning that the liquid will probably be water. For simplicity, we would probably want to make the polymer in the water and then use the product mixture directly without isolation. This process, called *emulsion polymerization*, sounds almost too good to be true.

As discussed in Chapter 4, emulsion polymerization received a significant boost in the United States during the Second World War. When Japan overran countries that supplied natural rubber to the West, a crash program to manufacture synthetic rubber was initiated in the United States and Canada. The product was called Government Rubber-Styrene (GR-S), and was produced by the emulsion polymerization of butadiene and styrene. The fundamental recipe for GR-S is still used as a teaching tool for those learning the art and science of emulsion polymerization.

For an emulsion polymerization, monomer is added to an aqueous mixture containing a surfactant (detergent or soap) and a water-soluble initiator. (An *emulsion* is a colloidal suspension of tiny droplets of one liquid dispersed in another. Colloids have particle sizes between approximately 10 nm and 1 mm. Milk is a common example of an emulsion in which fat globules are suspended in water.) Surfactants generally have long hydrocarbon chains with a polar functional group at one end. In water, the surfactant molecules organize themselves into micelles, extremely small aggregates on the order of 10 nm in diameter. The hydrocarbon chains, being hydrophobic (water-hating) mix together inside the micelles. The water-soluble, polar functional groups (hydrophilic or water-loving) on the ends of the chains form the outer surface of the micelles. (In washing our hands, dishes, or clothes, greasy or oily "dirt" is dissolved or suspended in the hydrophobic part of the micelle. The polar end groups keep the micelles suspended in water, allowing the dirt to be removed during rinsing.) The monomer, which is not very soluble in water, slowly diffuses from large monomer droplets through the water to the micelles. Initiating radicals are generated and migrate to the surface of the micelles, where they encounter monomers. Polymerization occurs inside the monomer-swollen micelles by essentially a bulk polymerization process. The micelles containing polymer molecules, now called polymer particles, continue to grow as fresh monomer diffuses in constantly. The rates of emulsion polymerizations are very high, and reactions are typically run at very high monomer concentrations (25% to 60% by mass). The result is a white emulsion resembling milk, called a latex. Latexes have low viscosities and are often used directly without purification or isolation. The process is summarized in Figure 5-3.

Figure 5-3. Basic species present in an emulsion polymerization.

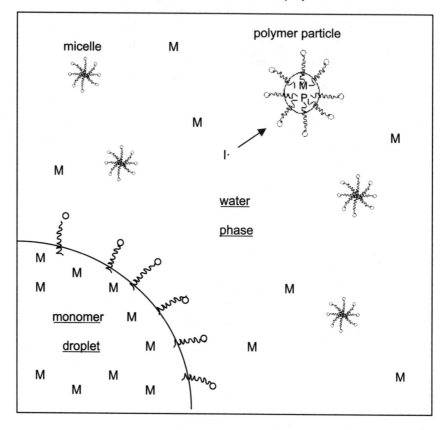

Key

I·	radical initiator (migrating toward particle surface)
M	monomer
P	polymer
ᴊᴜᴠᴠᴏ	surfactant molecule: circle represents polar functional group; squiggly line represents long alkyl chain

Latexes find wide use in industrial and consumer applications, particularly in the areas of synthetic rubber and coatings. Latex paints, for example, provide excellent alternatives to older, oil-based (alkyd) paints, which release substantial quantities of organic solvents to the atmosphere as they dry. Because volatile organic compounds (VOCs) are a factor in depleting the ozone layer, many industrial nations are striving to develop products that have greatly reduced VOCs. The major volatile emission from latex paints is water. Other applications for latexes include gloves, rubber tubing, adhesives, coatings for paper, carpet backings, toughened plastics, floor polishes, sealants, and drug-delivery materials. Some types of synthetic rubber are still manufactured by emulsion polymerization, especially styrene-butadiene rubber (SBR).

Copolymerization

So far in our tour of radical chain-growth reactions, we have discussed mainly homopolymerizations. Sometimes (often) to obtain a polymer with the properties needed, it would be desirable to have a random copolymer. One monomer might be chosen because of its thermal properties while a second component is added to improve adhesion. By making copolymers, one can produce an extensive assortment of polymers with a wide range of properties. Consider polystyrene, by itself a rather brittle polymer. If styrene is copolymerized with the monomer acrylonitrile, the resulting copolymer (SAN) is much less brittle (has a much higher impact strength; see Chapter 7). A copolymer of styrene and butadiene is actually elastic. A terpolymer of all three monomers (ABS) results in very useful materials with toughness and elasticity. Table 5-2 summarizes typical applications.

Table 5-2. Styrene and some of its copolymers.

Monomers	Polymer Abbreviation	Properties	Typical Uses
styrene	PS	clear, brittle	packaging, toys, disposable beverage containers, and tableware
styrene + acrylonitrile	SAN	less brittle than PS, tougher	battery cases, furniture, kitchenware
styrene + butadiene	SBR	rubbery, tough	tires, shoe soles, flooring, electrical insulation
styrene + acrylonitrile + butadiene	ABS	very tough	appliances, snowmobiles, automotive parts, luggage, telephones

To make such materials, what might happen if we simply mixed two (or more) monomers in a polymerization reactor? Many pairs of monomers participate in copolymerizations in a straightforward way. In other cases, however, one monomer prefers to polymerize with itself rather than with the other monomer. In these cases the reaction mixture at the end of the polymerization contains polymer samples with widely different compositions rather than the expected random copolymer. Some chains are much richer in one monomer while others are much richer in the other monomer. Although sometimes this heterogeneity is a problem, many times the polymer mixture performs its function quite well. Both quantitative methods and experimental techniques exist that help produce more random copolymers when necessary.

Rather than worry about whether a copolymer of two different monomers is heterogeneous or not, why not just make homopolymers of each and mix them together (make a *polymer blend*)? The answer is easy: Most polymers do not "like" other polymers, and, on mixing, each will tend to segregate into a phase that excludes any other polymers (will *phase separate*). These different phases tend not to stick to each other very well, resulting in a polymer sample with poor physical properties. We will discover the practical consequences of this in Chapter 7 when we discuss the properties of certain block copolymers called thermoplastic elastomers and in Chapter 9 when we discuss the recycling of plastics.

What's Up with Super Glue?

Super glue, that instant adhesive used for so many jobs around the house, illustrates ionic chain polymerization beautifully. The glue contains a cyanoacrylate monomer that polymerizes by an anionic mechanism. Even a trace of water or OH group (for example, that found on your skin) causes an extremely rapid polymerization forming a rigid plastic bonded to the surface(s) on which it polymerized.

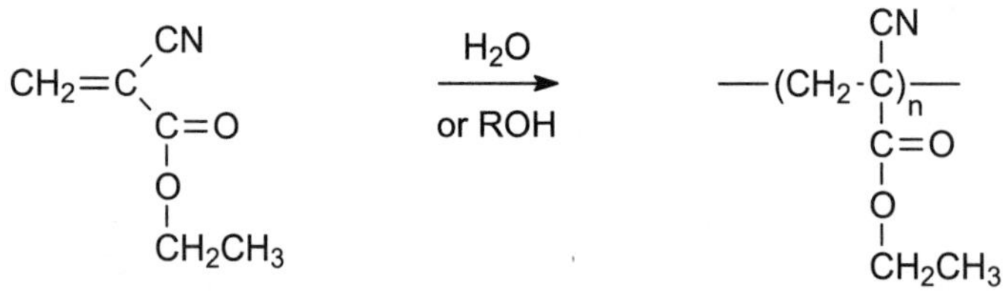

$$CH_2=C \begin{matrix} CN \\ \\ C=O \\ | \\ O \\ | \\ CH_2CH_3 \end{matrix} \quad \xrightarrow[\text{or ROH}]{H_2O} \quad -(CH_2\text{-}C)_n- \begin{matrix} CN \\ | \\ \\ | \\ C=O \\ | \\ O \\ | \\ CH_2CH_3 \end{matrix}$$

ethyl cyanoacrylate   poly(ethyl cyanoacrylate)

The monomer was synthesized in the early 1950s at the Tennessee Eastman Company in an effort to produce an extremely tough acrylic polymer for jet airplane canopies. Chemists had a very difficult time analyzing the monomer because it polymerized so rapidly. It never found use in jet aircraft.

But the chemists at Eastman recognized that a monomer such as this would be very useful as an adhesive, including one for closing medical wounds. In the early 1960s they applied for approval from the U.S. Food and Drug Administration (FDA). The U.S. military learned of the material and quickly sent it to Vietnam. Medics in the field were able to stop the bleeding in severe battle wounds by simply spraying the affected areas with cyanoacrylate monomer, saving countless lives.

Subsequently, cyanoacrylates have been used since the 1970s in Europe for a variety of surgical procedures and more recently in Canada. As a topical skin adhesive, cyanoacrylate enables emergency rooms to close wounds faster and reduces the need for stitches. Although veterinarians have been using cyanoacrylates for years, the adhesive was not approved by the FDA until 1998 for use in the United States on humans.

Ionic Chain Polymerization

Radical polymerization, while the most common mechanism for chain-growth polymerization, is not the only technique for forming long chains of carbon-carbon bonds. We mentioned earlier that both cations and anions are possible reactive intermediates. Forming a positive or a negative charge on carbon gives rise to intermediates that are substantially more reactive than carbon free radicals, however. This means that even small amounts of impurities such as water or alcohol will terminate an ionic polymerization, resulting in very low molar mass product or no product at all. However, if we could reduce the levels of these impurities to extremely low levels, termination reactions would essentially disappear, opening up the possibility for a "living" polymerization. More about that later. Note that termination by coupling is not possible with cationic or anionic intermediates, because species with like charges repel each other. Disproportionation reactions are also rare. Ionic polymerizations are always carried out in a solvent system, and because of the high reactivity of the intermediates, often at low temperature. Compared to radical reactions, which take hours to days to reach completion, ionic polymerizations are usually extremely fast. Chemists who specialize in anionic polymerization might spend most of a week cleaning glassware and purifying solvents and monomers for a polymerization that might be over in a few minutes.

To begin, let's consider the *anionic polymerization* of styrene. For an initiator, we will choose an *organometallic compound* (an organic compound bonded to a metal atom) such as butyllithium, $C_4H_9^-$ Li^+. Although the details differ, you should recognize the overall similarity of the mechanism for this anionic polymerization to that for the free radical polymerization of ethylene, above (initiation, propagation, and termination).

Initiation involves butyllithium attacking a molecule of styrene, forming the initiating species, a styryl anion. Ordinarily this is an extremely rapid reaction. The styryl anion attacks another styrene molecule, beginning the propagation stage, which also proceeds quickly, even at low temperature. If the monomer and solvent are pure enough, there will be no termination, unlike radical chain polymerization. If the polymerization is carried out properly, all of the styryl anions are generated at the same time, and they then consume all of the monomer at the same rate. At this point, the reaction mixture consists of a number of polymer chains, all of approximately equal length (equal degree of polymerization) and all terminated as a styryl anion.

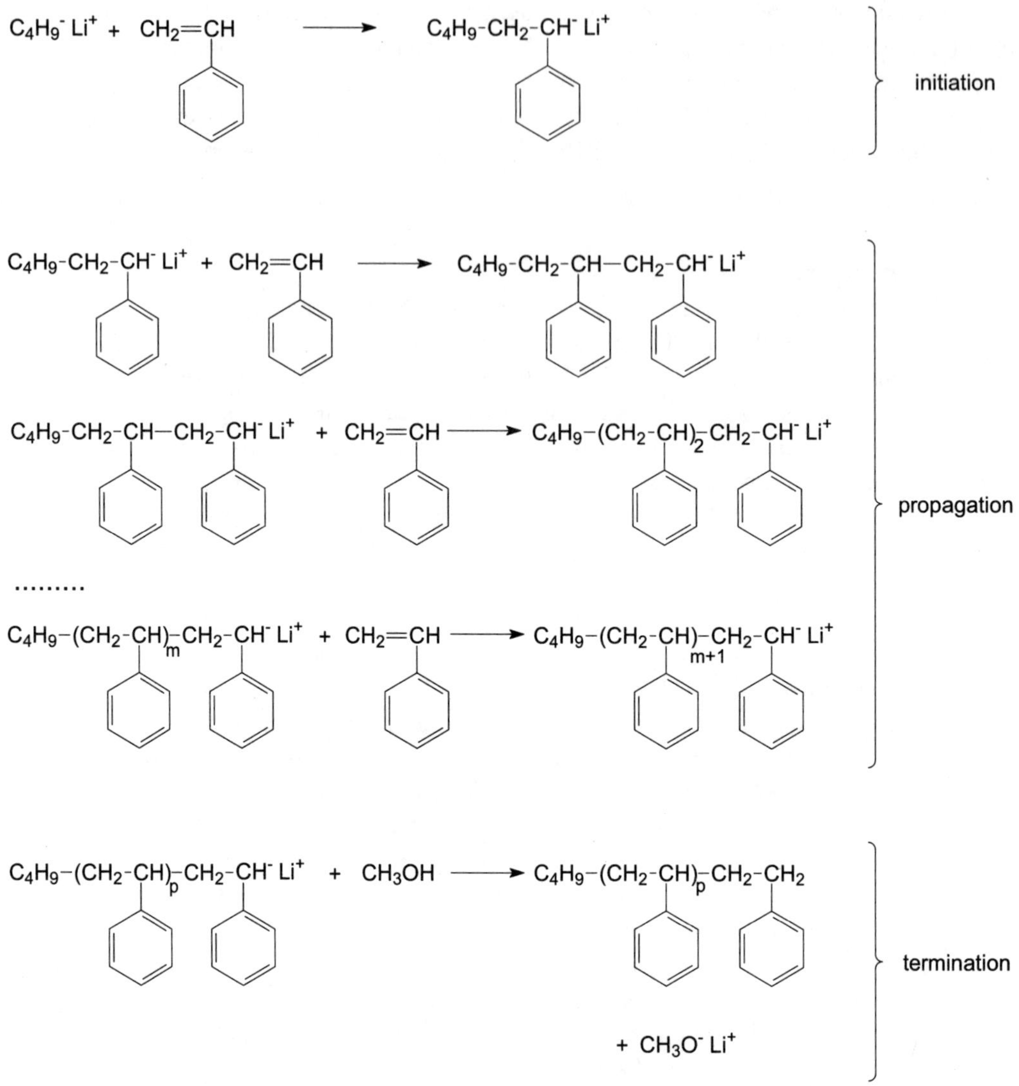

$C_4H_9^- Li^+ + CH_2=CH \longrightarrow C_4H_9-CH_2-CH^- Li^+$

initiation

$C_4H_9-CH_2-CH^- Li^+ + CH_2=CH \longrightarrow C_4H_9-CH_2-CH—CH_2-CH^- Li^+$

$C_4H_9-CH_2-CH—CH_2-CH^- Li^+ + CH_2=CH \longrightarrow C_4H_9-(CH_2-CH)_2-CH_2-CH^- Li^+$

....

$C_4H_9-(CH_2-CH)_m-CH_2-CH^- Li^+ + CH_2=CH \longrightarrow C_4H_9-(CH_2-CH)-CH_2-CH^- Li^+$ m+1

propagation

$C_4H_9-(CH_2-CH)_p-CH_2-CH^- Li^+ + CH_3OH \longrightarrow C_4H_9-(CH_2-CH)_p-CH_2-CH_2$

$+ CH_3O^- Li^+$

termination

It Lives!

Now one of two things can occur. A chain terminator can be added (e.g., a small amount of an alcohol) from which each chain will abstract a hydrogen ion and become a neutral polymer (a dead chain). This is the termination step shown in the preceding reaction sequence. Or, more monomer can be added and the polymerization will continue until the new sample of monomer has been consumed. In other words, these anionic polymerizations are "living" polymerizations, so named because the chains remain active until they are deliberately terminated (become "dead"). (The terms "living" and "dead" describe relative states of chemical reactivity only and not any biological condition.) Such behavior does not happen during normal free radical polymerizations, where termination reactions occur during the course of the polymerization and compete with initiation and propagation.

Block Copolymerization

Instead of adding more styrene to the polymerization above, what if we were to add a second monomer, say methyl methacrylate, after the first monomer had been totally consumed? The result would be an AB *block copolymer*. "A" represents the first monomer, styrene in our example, and "B" the methyl methacrylate. The degrees of polymerization of the two blocks could be the same, or they could be different. The steps involved in constructing the methyl methacrylate block are shown in the following scheme:

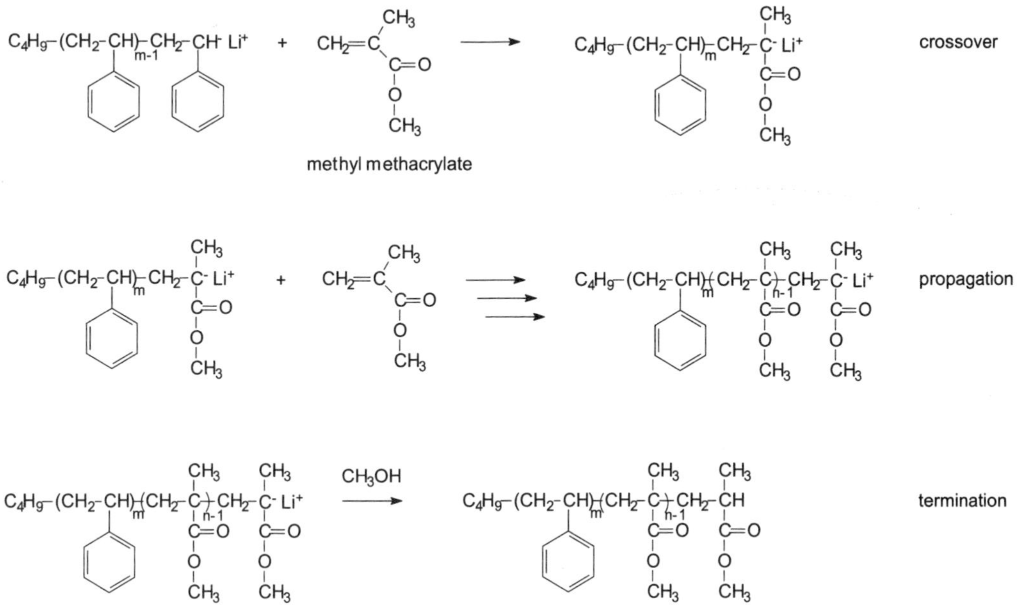

methyl methacrylate

crossover

propagation

termination

The reactions are just like the ones that produce polystyrene. First, a molecule of methyl methacrylate reacts with a polystyryl anion to form a methyl methacrylate anion (crossover). Then the methyl methacrylate block grows in a series of propagation steps, followed by the addition of methanol to effect termination as before. We can draw the structure for the block copolymer, named poly(styrene-*block*-methyl methacrylate) as is shown here.

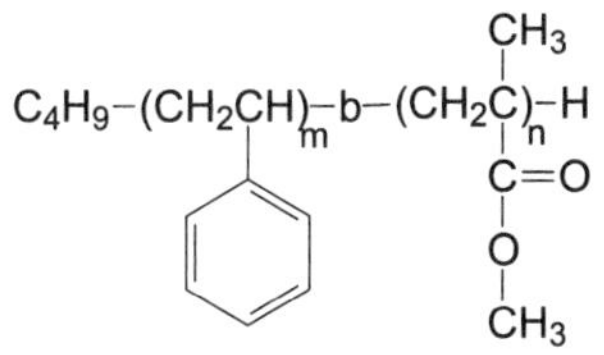

The structure indicates that styrene was polymerized first, initiated with butyllithium, followed by the methyl methacrylate. The "b" is included in the structure to confirm that this is indeed a block copolymer, not a random copolymer. Using similar chemistry, it is possible to synthesize *triblock* copolymers. Usually the two end blocks are the same (prepared from monomer "A") while the center block is different ("B" monomer). Such a polymer would be designated A-b-B-b-A, or more simply ABA.

It is important to appreciate that polymer produced by an anionic chain-growth mechanism can have drastically different properties from one made by a normal free radical reaction. Block copolymers can be synthesized in which each block has different properties. We mentioned in Chapter 4 that Michael Szwarc of Syracuse University developed this chemistry in the 1950s. Since that time, block copolymers produced by anionic polymerization have been commercialized, such as styrene-isoprene-styrene and styrene-butadiene-styrene triblock copolymers (e.g., Kraton from Shell Chemical Company). They find use as *thermoplastic elastomers (TPE)*, polymers that act as elastomers at normal temperatures but which can be molded like thermoplastics when heated. We will discuss TPEs further in Chapter 7.

Let's consider some of the implications of these living polymerizations. Because all of the chains are initiated simultaneously at the start of the polymerization, all of the chains of a homopolymer will have approximately the same degree of polymerization (DP). For the first time, we can synthesize a polymer where all of the molecules are the same. Amazing! Actually, they are not all exactly the same, but if the synthesis is carried out with sufficient skill, the polymer sample will have in fact a very small molar mass distribution (will have very low polydispersity). Without termination, the molar mass depends upon only the ratio of monomer concentration to initiator concentration. Therefore, one can *design* the synthesis to produce polymer of a particular molar mass. And unlike that for free radical polymerization, the molar mass for an ionic polymerization is directly proportional to percent conversion, as shown in Figure 5-4. For reference, the curves for radical chain-growth and step-growth polymerizations (Figure 5-2) are included in Figure 5-4.

Cationic Polymerization

Some monomers are also polymerized by a *cationic* mechanism in a series of steps not too unlike those of anionic chain-growth. Initiators are often Lewis acids such as $AlCl_3$. The polymerization is not quite as straightforward as anionic, because for one thing cationic intermediates are subject to more side reactions. Common monomers that undergo cationic polymerization include styrene, isobutylene, and vinyl acetate. Some commercial products

Figure 5-4. Polymer molar mass versus polymer formation for living polymerization.

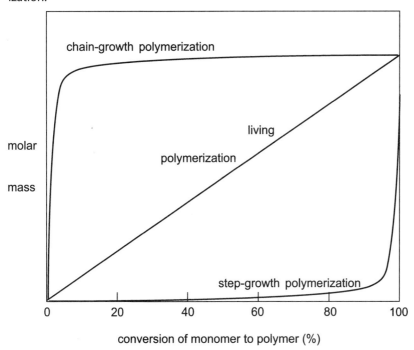

are prepared by cationic polymerization in large amounts. An example is butyl rubber, a copolymer of 2-methylpropene with a small amount of isoprene. Butyl rubber remains flexible at temperatures as low as −50°C and is more resistant to ozone than is natural rubber.

Polymer Stereochemistry: Tacticity

Recall that we introduced the idea of stereoisomers in Chapter 3 in the section on natural rubber. The stereochemistry in polyisoprene arises because of the rigid nature of the carbon-carbon double bond. Natural rubber is cis-polyisoprene while gutta percha is trans-polyisoprene, the two polymers having drastically different properties

We can also have stereoisomers in compounds without carbon-carbon double bonds. For example, the relative positions of groups along a polymeric chain of carbon atoms can give rise to different stereoisomers, a situation we call *tacticity*. To understand this, let's consider some different ways of drawing a segment of a polymer chain of polypropylene:

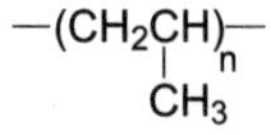

$$-(CH_2CH)_n-$$
$$CH_3$$

A

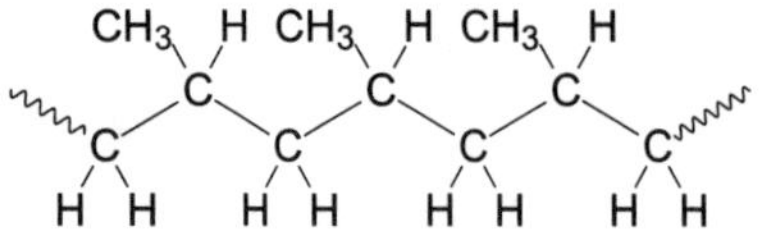

B

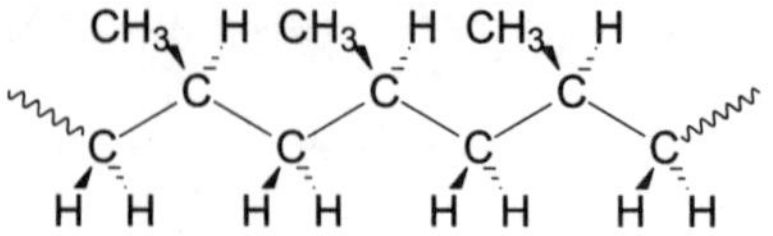

C isotactic

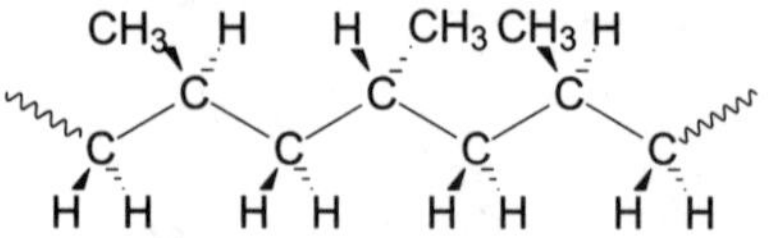

D syndiotactic

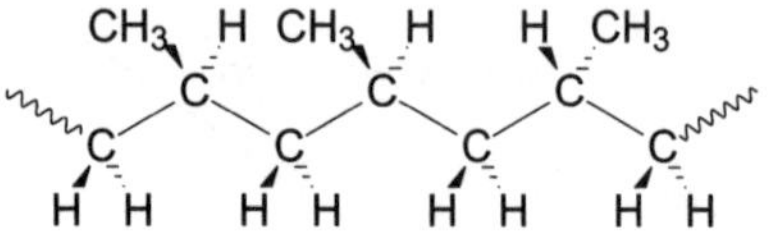

E atactic

Structure A is the representation we are used to and only shows us the repeat unit. Structure B shows part of the polymer backbone with the hydrogens and CH_3 groups attached. The squiggly lines at each end simply indicate that the chain continues in each direction. Three-dimensional arrangements of the various atoms are not implied. We know, however, that a saturated carbon atom has tetrahedral, not planar geometry. Structures C, D, and E attempt to illustrate this. We show the polymer chain aligned in a planar, zigzag line lying on the page. The groups bonded to the chain at the end of the solid, wedge-shaped bonds are defined as being in front of the plane of the paper. The groups bonded to the broken line bonds are behind the plane. Structures C, D, and E depict the three possible stereochemical arrangements for the CH_3 groups in polypropylene (three stereoisomers). In C all of the CH_3 groups are lined up on one side of the carbon atoms. This isomer is called *isotactic*. In D, the CH_3 groups alternate between in front, behind, in front, and so on. This isomer is called *syndiotactic*. Finally, the relative arrangements of the CH_3 groups can be random, neither all on one side nor alternating. This configuration, shown in structure E, is called *atactic*.

Why is this important? As we pointed out in Chapter 4, atactic polypropylene is a soft, amorphous polymer with poor physical and thermal properties that is not a very useful material. Isotactic polypropylene, on the other hand, is a crystalline polymer with a high melting point (183°C) and excellent physical properties. This is the reason that Giulio Natta's preparation of isotactic polypropylene was so significant. The regularity along the chain allows the chains to pack together more efficiently, increasing crystallinity. As we will see in Chapter 7, crystalline polymers generally not only have better thermal and mechanical properties, but also have better resistance to attack by solvents and other chemicals. These properties are very important in applications such as automobile parts.

This leads to another important question—how do we control the tacticity of polymers? We cannot change from one configuration to another by rotating groups around bonds. To do so would require that a bond to a carbon atom must be broken and then reformed. Once we have made a polymer that is, say, atactic, we can't somehow make it become isotactic. As some of the pioneers in polymer synthesis surmised during the middle of the twentieth century, tacticity must develop when the polymer is made. Most radical chain-growth polymerizations produce polymer with random (or atactic) stereochemistry. Such a preparation would produce atactic polypropylene. Isotactic or syndiotactic polymers are usually prepared using metallic or metal-organic catalysts (e.g., by anionic or coordination polymerization).

Coordination Polymerization

The line between ionic and *coordination polymerization* is a bit blurry. It was discovered that the tacticity one obtained during polymerization could be controlled by changing the anionic initiator. As we discussed in Chapter 4, Karl Ziegler, Natta, and several other twentieth-century chemists discovered a large number of catalyst systems for polymerizing ethylene, propylene, and a variety of dienes. When Ziegler began studying the reactions of alkenes with organolithium and various aluminum hydride compounds, he discovered that adding small amounts of certain transition metal compounds caused dramatic changes in the outcome of these reactions. These catalyst systems, called *coordination catalysts*, are responsible primarily for producing one tacticity in preference to the others in alkene polymerizations. The consequences, as we mentioned, are extremely significant, and earned these two men a Nobel Prize in 1963. Many of these polymerizations occur in part because the electrons of the double bond in the monomer coordinate or complex with a transition metal in the catalyst, influencing the rate of the reaction as well as the stereochemistry during polymerization.

Ethylene can be polymerized at high temperatures and pressures using free radicals, as we have seen. The product is highly branched, primarily because of the high temperatures used in the polymerization. Ziegler discovered that polymerizing ethylene with what we now call coordination catalysts produces a different polyethylene, one that is denser, less highly branched, and more highly crystalline. This is called high-density polyethylene (HDPE), while the free radical product is referred to as low-density polyethylene (LDPE). Isotactic polypropylene is produced as well using coordination catalysts, also known as *Ziegler-Natta catalysts*.

Coordination Polymerization of 1,3-Dienes

We have not discussed the polymerization of dienes very much so far. Part of the reason for this stems from the added complexity caused by the second double bond. Depending upon the polymerization conditions chosen, polymers with mixtures of repeat units can form. If only one of the double bonds participates in the polymerization, this is referred to as 1,2-polymerization. If both double bonds participate, the polymerization is called 1,4. Common monomers that fall in this category include 1,3-butadiene and isoprene:

$$CH_2{=}CH{-}CH{=}CH_2 \qquad\qquad CH_2{=}\underset{\underset{CH_3}{|}}{C}{-}CH{=}CH_2$$

1,3-butadiene   isoprene

The products are elastomers (recall that the starting material for natural rubber is isoprene). Butadiene can give almost complete 1,2- or 1,4-polymer, depending primarily upon the coordination catalyst and the polymerization conditions:

$$-(CH_2\text{-}\underset{\underset{\underset{CH_2}{\|}}{\underset{CH}{|}}}{CH})_n{-} \qquad\qquad -(CH_2\text{-}CH{=}CH{-}CH_2)_n{-}$$

1,2   1,4

We have seen that the double bonds in a chain of polyisoprene can exist as cis and trans stereoisomers. *Synthetic* polyisoprene then has the added complexity of 1,2- versus 1,4-polymerization in addition to the possible existence of different stereoisomers about the double bond. As with butadiene, different coordination catalysts produce isoprene polymers with a preponderance of 1,2- or 1,4- polymer as well as different stereochemistry. Not unexpectedly, these different polymers possess strikingly different physical properties.

Single-Site (or Metallocene) Catalysts

Within the last decade or so, the development and increased understanding of catalysts have resulted in techniques to prepare specific polymers with targeted properties. A new type of transition metal catalyst has gained importance, one in which each metal atom has a single polymerization site. Ziegler-Natta catalysts, on the other hand, have several different polymerization sites. The use of these *single-site catalysts* (sometimes also called *metallocenes*) has led to new polymer families that were previously unknown or to polymers that were too difficult to synthesize using older technology. Many of these polymers have been commercialized in a very short time, revolutionizing the polyolefin industry. From the mid-1940s until near the end of the twentieth century, five basic polyolefins were commercialized: low-density polyethylene (LDPE), high-density polyethylene (HDPE), linear low-density polyethylene (LLDPE), polypropylene (PP), and ethylene-propylene-diene terpolymer (EPDM). Since 1990, at least 11 new polyolefin families have emerged, in large part the result of new catalyst systems (Sinclair 2001). In addition, single-site catalysts have enabled huge productivity gains in the manufacture of traditional polyolefins such as polyethylene. Polymers prepared from these new catalysts have found many new applications, such as polyethylene salad packaging with extended shelf life to name one example. The specificity of the catalysts enables the synthesis of polymers that have more regular structures and that have much improved physical properties. As a result, polymers prepared from inexpensive monomers such as ethylene and styrene are replacing much more expensive, specialty polymers in medical, electronic, and automotive applications. It is entirely possible that single-site catalyst technology will have a greater financial impact on the polymer industry than any other single technological innovation (Sinclair 2001).

Living Radical Polymerizations

So far we have discovered very few polymerization techniques for making macromolecules with narrow molar mass distributions and for preparing di- and triblock copolymers. These types of polymers are usually made by anionic or cationic techniques, which require special equipment, ultrapure reagents, and low temperatures. In contrast, most of the commodity polymers in the world such as LDPE, poly(methyl methacrylate), polystyrene, poly(vinyl chloride), vinyl latexes, and so on are prepared by free radical chain polymerization. Free radical polymerizations are relatively safe and easy to perform, even on very large scales, tolerate a wide variety of solvents, including water, and are suitable for a large number of monomers. However, most free radical polymerizations are unsuitable for preparing block copolymers or polymers with narrow molar mass distributions.

What would it take to achieve better control over radical polymerization so that, for example, block copolymers could be prepared? Remember that the key to making block copolymers anionically is the living nature of the intermediate—chain termination does not compete with initiation and propagation. Could we design a free radical system in which we could "turn off" termination reactions until we wanted them?

Within the last decade, techniques have been developed that begin to make this possible. They use the concept of trapping the intermediate free radicals and forming dormant species that are in equilibrium with the reactive, growing free radicals. At any given time, the concentration of active free radicals is extremely low, decreasing the possibility of two radicals finding each other and reacting. Thus, termination reactions are all but eliminated, and one has control over the chain reactions. This process is not as susceptible to impurities as an anionic or cationic polymerization reaction and is therefore easier to carry out. A typical process is shown in the reaction scheme below for styrene.

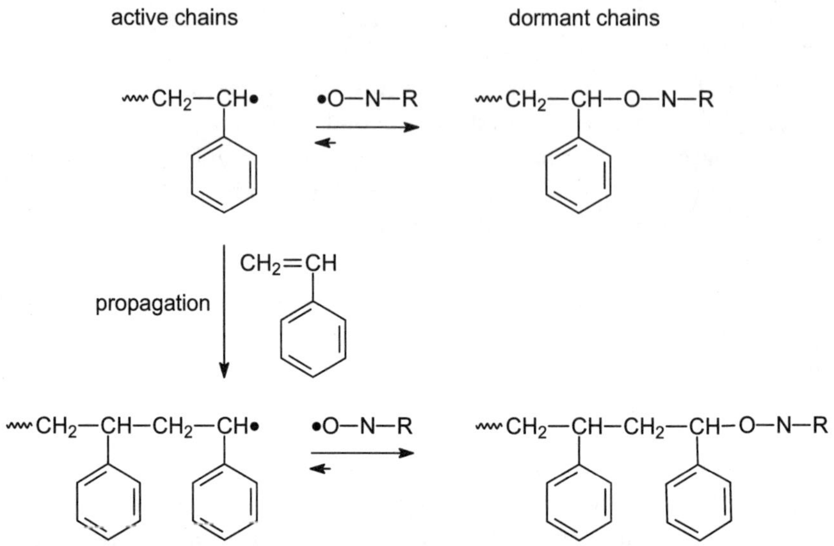

active chains   dormant chains

The species ·O-N-R in the scheme is a stable organic radical called a nitroxide, one type of radical that does not react with itself but which reacts with carbon radicals forming weak C-O-N bonds. This approach continues to develop and allows the synthesis of polymers with very narrow molar mass distributions, block copolymers, and polymers with different architectures such as highly branched materials (discussed shortly). This is but one of a limited number of techniques that show promise for producing improved thermoplastics, elastomers, and adhesives for packaging and automotive applications (Anon. 2002).

Other Types of Polymerizations, Polymers
Ring-Opening Polymerization
Some cyclic compounds can undergo ring-opening polymerization, and some of these polymers are commercially very important. Representative ring compounds and a few specific examples are summarized in Table 5-3. Poly (ethylene oxide) (PEO) and polyethyleneimine are both water-soluble polymers of commercial importance. We'll explore these materials further in Chapter 6 when we talk about solution properties. Because the three-membered ring in ethylene oxide is highly strained, it undergoes ring-opening quite easily, and polymerizes with either acidic or basic initiators. Some of the reactions are living polymerizations, meaning that block copolymers can

Table 5-3. Examples of cyclic compounds that undergo ring-opening polymerization.

Monomer	Structure	Polymer Name	Polymer Structure
Cyclic ether		**Polyether**	
ethylene oxide	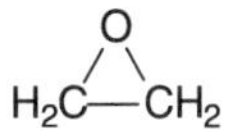	poly(ethylene oxide)	$-(CH_2-CH_2-O)_n-$
tetrahydrofuran	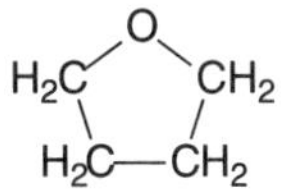	polytetrahydrofuran	$-(CH_2CH_2CH_2CH_2-O)_n-$
Lactone (cyclic ester)		**Polyester**	
δ-valerolactone	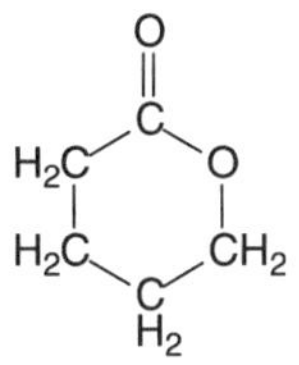	poly(valeric acid)	$-(O-CH_2CH_2CH_2CH_2C)_n-$
Cyclic amine		**Polyamine**	
aziridine	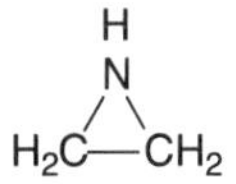	polyethyleneimine	$-(CH_2CH_2-NH)_n-$
Lactam (cyclic amide)		**Polyamide**	
ε-caprolactam	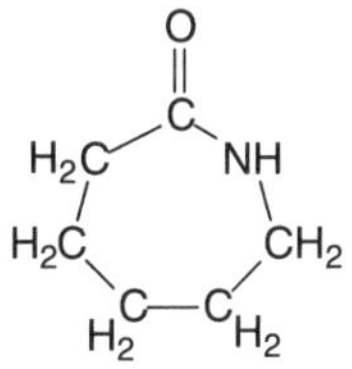	nylon-6	$-(NH-CH_2CH_2CH_2CH_2CH_2C)_n-$
Cyclic alkene		**Polyene or polycyclic**	
cyclobutene	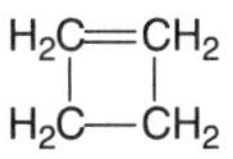	polybutadiene	$-(CH_2-CH=CH-CH_2)_n-$

be formed. Block copolymers in which the water-soluble PEO block is paired with an oil-soluble (hydrophobic) block have very useful properties and are available commercially.

Note that the polymerizations of lactones and lactams give polyesters and polyamides, respectively, polymers that can also be synthesized by conventional step-growth polymerization from noncyclic monomers. Although the polymer prepared from cyclic monomers looks like a condensation polymer, in strict terms it is not, because no small-molecule byproduct was produced in the polymerization. It is because of instances such as this that no one (or even two) scheme can completely classify polymers or the mechanisms of their preparation.

Ring-Opening Olefin Metathesis Polymerization (ROMP)

In the mid to late 1980s, transition-metal catalysts were developed that were particularly useful for carrying out *olefin* (alkene) *metathesis* reactions (Rouhi 2002). Metathesis is a reaction in which two molecules containing carbon-carbon double bonds exchange carbon atoms along with any groups attached to them:

Equation 13

$$R-CH=CH_2 \; + \; R'-CH=CH_2 \longrightarrow R-CH=CH-R' \; + \; CH_2=CH_2$$

"R" and "R'" refer to organic groups that contain some number of carbon atoms and may or may not be different from each other. With cyclic alkenes, metathesis leads to polymerization (specifically, *ring-opening metathesis polymerization* [*ROMP*]), shown here for cyclopentene:

Equation 14

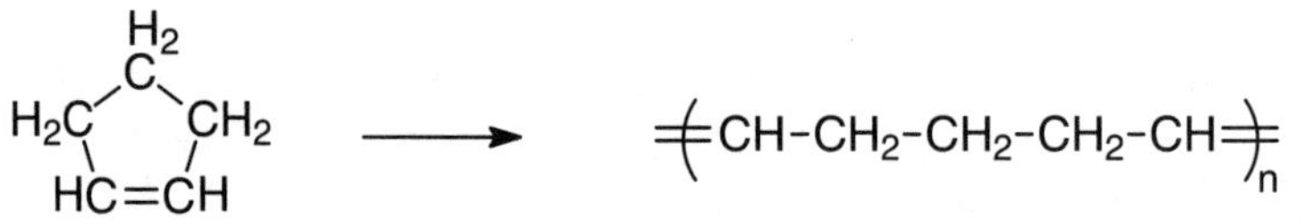

$$\underset{HC=CH}{\overset{\overset{H_2}{C}}{H_2C \diagdown CH_2}} \longrightarrow \left(CH-CH_2-CH_2-CH_2-CH \right)_n$$

Both olefin metathesis and ROMP proceed by the same mechanism, in which a transition metal catalyst reacts with the olefin to form an intermediate complex. People developing ROMP use computer molecular modeling to design transition metal catalysts that polymerize a wide variety of cyclic olefins in specific ways. New, more sophisticated catalysts continue to be developed, many from the laboratories of Robert Grubbs at the California Institute of Technology and Richard Schrock at the Massachusetts Institute of Technology. Monomers with two double bonds can be utilized to make crosslinked materials. Using a process called *reaction injection molding* (see Chapter 8), polymerization and crosslinking can take place almost simultaneously, producing an automobile body part, a satellite antenna dish, or a snowmobile body, say, in one step.

Dendrimers: Polymers That Look Like Trees

Up until now we have mostly described macromolecules as being composed of long, linear chains. Until about 20 years ago, that was the structure that polymer scientists almost always strove to synthesize, to utilize, and to understand. Although the covalent bonds that make up the chains are essentially one-dimensional, most macromolecules assume three-dimensional shapes such as random coils, as we discuss in Chapters 6 and 7. We have mentioned only briefly other structures or architectures such as branched, graft, or crosslinked polymers. Branched and graft polymers are usually prepared by a free radical process. Because of this, the branches or grafts occur randomly along the central chain and vary in length. For many applications, the heterogeneity of these molecules has no adverse affect on the usefulness of the material. For some applications, however, it would be desirable to be able to control the number of branches, their length, and their placement in the macromolecule. Some of the newer synthetic techniques reduce some, but not all, of the heterogeneity.

Carrying the branching idea to the extreme, what if we could make a molecule that was so extensively branched that it had no main chain? In 1984 Donald Tomalia at Dow Chemical disclosed symmetrical structures with controlled branching that he called *dendrimers*. The name is derived from the Greek words for tree (*dendron*) and part (*mer*). These macromolecules have highly ordered structures, are globular in shape, and usually have reactive groups on the outside. They are built up in a series of repetitive synthetic steps, each step adding a layer or *generation* to the structure. This is shown schematically in Figure 5-5. At the center is the core (illustrated with a triangle) to which is bonded in this case three functional groups (Z). In our hypothetical example, the monomer is:

Here, Z reacts with Y to form a chemical group represented by a square. Therefore one molecule of core reacts with three molecules of monomer to form the first generation, which contains 6 Z groups on the outside. One molecule of first generation dendrimer then reacts with 6 molecules of monomer to form the second generation molecule. In the second generation there are 12 Zs, and, in the third, 24. The functional groups on the outside of the dendrimer, as well as the number of monomer units, increase in an exponential relationship with each successive generation. Eventually (probably by the 5th to the 8th generation depending upon the monomer), the reactive groups in the dendrimer become too crowded to react with any more monomer. These molecules take on a globular shape, with the end groups virtually covering the surface. Typical sizes of dendrimers are in the range of 3 to 10 nm (Simanek and Gonzalez 2002). So here is a truly unusual macromolecule, one whose properties are dominated by the end groups rather than by the repeating units making up the bulk of the polymer. In contrast, linear polymers have only two end groups. When the degree of polymerization is high, the effect of these two end groups is negligible.

Figure 5-5. The first three generations of a hypothetical dendrimer built up from a trifunctional core.

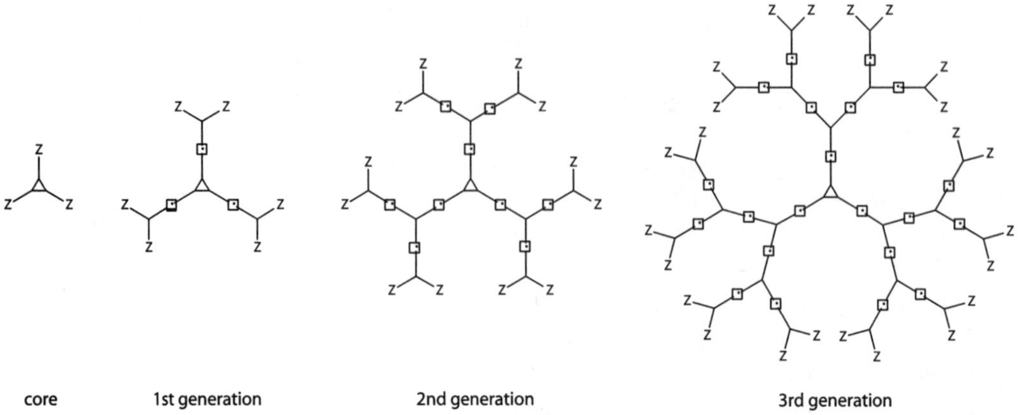

core   1st generation   2nd generation   3rd generation

You may have already noticed a potential problem with the hypothetical reaction we outlined above, where we said that Y reacts with Z to form something else (represented by a square in Figure 5-5). If the monomer also contained Z groups, molecules of monomer would react with each other as well as with the core to form a mixture of highly branched macromolecules. (These compounds, called *hyperbranched polymers*, have irregular, more random branching and therefore lack the symmetry of the perfect dendrimer. However, they are much easier to synthesize and their properties resemble those of dendrimers, making hyperbranched polymers attractive candidates for certain commercial applications [Turner and Voit 1997].) To carry out the chemistry leading to a perfect dendrimer, the Z groups on the monomer must be *blocked* or *protected*. In other words, they must be temporarily converted into something else that does not react with Y. After the first generation is formed, the blocking or protecting group is removed, regenerating Z again. These react with more monomer to form the 2nd generation. These steps—reaction, purification, deblocking—are repeated until the final dendrimer is obtained. Needless to say, this is tedious work, making dendrimers rather expensive specialty polymers.

In practice, dendrimers are either prepared as above, beginning with the core and building out, or constructed from the outside in. Tomalia developed the former method, which is used to make some commercial material (Tomalia and Fréchet 2002). In the latter approach, pioneered by Jean Fréchet and Craig Hawker in about 1990, a portion or "wedge" of the final dendrimer is prepared with one reactive functional group at the innermost point. In the final steps, three, four, or many wedges react with the core, forming the final dendrimer (Hecht and Fréchet 2001). Using this approach, it is possible to construct a dendrimer with two or more chemically distinct wedges, each with its own functional groups. This is illustrated in Figure 5-6, in which

Figure 5-6. Schematic construction of a dendrimer composed of two chemically different wedges or dendrons.

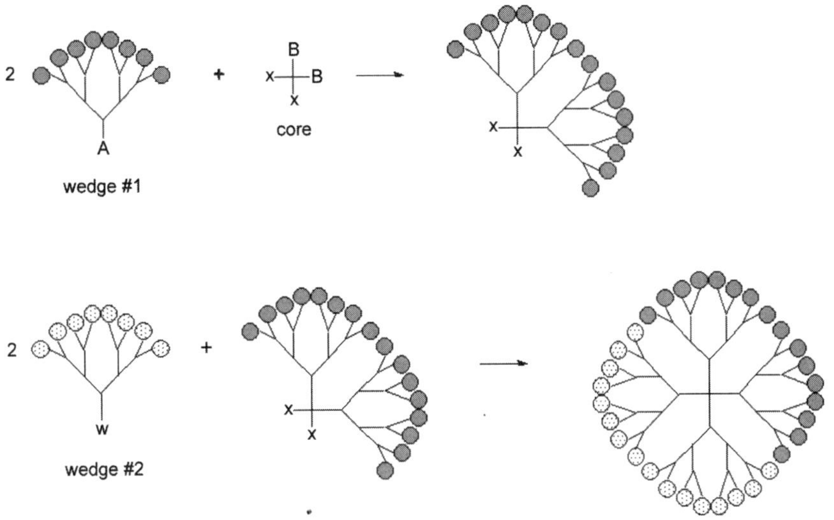

two wedges with "gray" functionality on the outside react with two arms of a four-armed core (A + B). This half of a dendrimer then reacts with two molecules of a different wedge (with "speckled" functionality), forming the final dendrimer (W + X). The shape of a dendrimer is largely controlled by the nature of the core. Cores based on small, symmetrical molecules tend to result in roughly spherical dendrimers. A core that has rod-like structure (perhaps itself a polymer) will yield dendrimers with cylindrical shape.

What are dendrimers good for? Being globular molecules, dendrimers do not entangle very much with each other. Therefore, they tend to be quite soluble, forming solutions with relatively low viscosities (see Chapter 6). Unlike the outside of the molecules, the core area tends to be quite uncongested. Small molecules can be encapsulated inside, providing compounds for drug or gene delivery. By choosing the proper core, one can construct, for example, a catalyst, a light-harvesting compound, or a light-emitting diode (LED) (Grayson and Fréchet 2001). This is a new field of polymer science, one that will continue to develop and provide a range of interesting and useful materials for the next several years. Examples will surely include nanomaterials and molecular devices (see Chapter 10).

Silicones

Polysiloxanes, or *silicones* as they are commonly called, are polymers of silicon, not carbon. Their chains are made up of alternating silicon and oxygen bonds and are characteristically very flexible (have very low glass transition temperatures—see Chapter 7). As a result, silicones typically find use as elastomers. Low molar mass compounds are used as lubricating oils. They are often referred to as inorganic polymers because the backbone contains no carbon atoms. However, organic groups are bonded to each silicone atom, the specific choice of group affecting the properties of the polymer. They can be synthesized either by condensation reactions or by ring-opening polymerization as illustrated below:

Equation 15

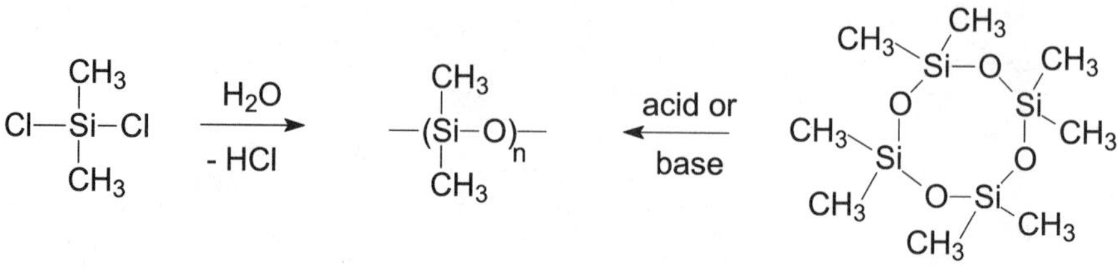

Because of their inherent flexibility, even at low temperatures, and resistance to oxidation, silicones are often used as sealants and caulking materials. For the latter application, they usually emerge from the tube as a viscous liquid so that they can be squeezed into an opening, whereupon they undergo a fairly rapid crosslinking to form a soft solid.

Summing Up

We have seen that polymers can be synthesized using a variety of chemistries. Monomers for chain-growth polymers usually contain a carbon-carbon double bond and are loosely called vinyl monomers. These monomers undergo chain-growth polymerization readily through free radical intermediates, historically the most prevalent method for polymerizing these materials. The reactions are carried out in bulk, in solution, or as a suspension (bead) or an emulsion (latex) in water. In contrast to free radical polymerization, increased control of structure can be obtained by using coordination catalysts, metal-organic materials such as Ziegler-Natta catalysts or the newer single-site catalysts. Linear polymers with decreased branching, stereoregularity (tacticity) and superior physical properties result. All of these processes are carried out on extremely large scales in continuous reactors producing multimillion tonnes of inexpensive polymers such as polyethylene (HDPE, LDPE, LLDPE), polypropylene (PP), poly(vinyl chloride) (PVC), and polystyrene (PS) each year. "Living" polymerizations using ionic catalysts or controlled free radical processes allow the preparation of well-defined polymers with highly controlled molar mass as well as unusual architectures (e.g., block copolymers).

Other polymers are prepared by a step-growth mechanism. Many of these are condensation reactions in which two different functional groups react, forming a new functional group (e.g., ester, amide) and a small-molecule byproduct. Often these reactions can be carried out "neat," meaning no solvents are used. These reactions account for millions of tonnes each

114

year of polyester for bottles and fiber, nylons for carpeting, and polyurethane foam for upholstered furniture and sneaker soles. Some cyclic compounds undergo ring-opening polymerization, forming a large number of commercially important polymers (e.g., epoxides, polyesters, polyamides, and silicones).

Although linear polymers make up the bulk of commercial materials, alternate architectures such as crosslinked materials (thermosets), branched or graft copolymers, or dendrimers and hyperbranched polymers are used because of their unusual properties.

References Cited

Anon. 2002. In control of a living process. *Chemical and Engineering News* September 9: 36–40.

Flory, P. J. 1953. *Principles of polymer chemistry*. Chapter 2. Ithaca, NY: Cornell University Press.

Grayson, S. M., and J. M. J. Fréchet. 2001. Convergent dendrons and dendrimers: From synthesis to applications. *Chemical Reviews* 101: 3819–67.

Hecht, S., and J. M. J. Fréchet. 2001. Dendritic encapsulation of function: Applying nature's site isolation principle from biomimetics to materials science. *Angewandte Chemie International Edition in English* 40: 74–91.

Rouhi, A. M. 2002. Olefin metathesis: Big-deal reaction. *Chemical and Engineering News* December 23: 29–33.

Simanek, E. E., and S. O. Gonzalez. 2002. Dendrimers: Branching out of polymer chemistry. *Journal of Chemical Education* 79 (10): 1222–31.

Sinclair, K. B. 2001. Future trends in polyolefin materials. *Macromolecular Symposia* 173: 237–61.

Tomalia, D. A., and J. M. J. Fréchet. 2002. Discovery of dendrimers and dendritic polymers: A brief historical perspective. *Journal of Polymer Science: Part A: Polymer Chemistry* 40: 2719 –28.

Turner, S. R., and B. I. Voit. 1997. Hyperbranched polymers. *Polymer News* 22: 197–202.

Other Reading

Nicholson, J. W. 1997. *The chemistry of polymers*. 2nd ed. London: The Royal Society of Chemistry.

Odian, G. 1991. *Principles of polymerization*. 3rd ed. New York, NY: Wiley-Interscience.

Ponomareva, V. T., and N. N. Likhacheva. 2002. Production of plastics based on metallocene catalysts. *International Polymer Science and Technology* 29(2): T10–T14.

Chapter 6

Polymer Solutions and Dispersions

S o far, we have spent a considerable amount of time discussing the strength of polymers and their unique physical and mechanical properties. However, some applications take excellent advantage of the interesting properties that polymers bring to solutions. Examples include paints, motor oils, and some of the products we put on our hair. In addition, as we saw in the last chapter, some polymers are synthesized in solution. In Chapter 8 we will encounter other polymers that are fabricated from solution into useful products. In this chapter, we will present some of the important properties of polymer solutions and develop a basic understanding of their origin. Some polymers would "like" to dissolve but can't. We'll try to understand why, and see how to take advantage of this. And, finally, we'll investigate some uses for polymers that are not actually dissolved in a solvent but rather are dispersed in a liquid. In a *dispersion*, particles of polymers or other substances are suspended in a liquid.

Introduction

Because most polymers are organic compounds, we would expect them to be soluble in a variety of solvents. Sugar dissolves in water and motor oil dissolves in gasoline (a hydrocarbon) because of positive interactions between solute and solvent molecules. Sugar molecules such as sucrose or dextrose are polar and contain OH groups that interact with water through hydrogen bonding. Nonsynthetic motor oil, a high boiling mixture of hydrocarbons, is nonpolar and is attracted to the nonpolar molecules of gasoline through van der Waals interactions. Based on the simple empirical rule that "like dissolves like," we would predict that most organic polymers would dissolve in solvents of similar polarity or character. Although in many cases we would be correct, the solution behavior of polymers is not as straightforward as that of small molecules. Let's try to understand why.

Polymer Solutions
Random Coils

In Chapter 5 we saw that some polymers could be prepared in solution. In this type of synthesis, the monomers and the polymer are all soluble in the solvent chosen. As the polymer chains form, solvent molecules surround all parts of the chains. But what do the "linear" chains of most polymers actually look like as they float around in solution? For thermodynamic energy reasons they are not straight, but instead are randomly arranged in shapes that are called *random coils*. In solution, the random coils are greatly expanded and are filled with solvent molecules. The concept of a random coil was developed by Herman Mark and colleagues in the 1920s. To isolate the polymer at the end of the polymerization, one can add the polymer solution to a large excess of a liquid that is not a solvent for the polymer (a *nonsolvent*) but in which the polymerization solvent dissolves. Nonsolvent molecules replace most of the solvent molecules surrounding the polymer, causing the coils to contract and the polymer to *precipitate* from the solution. This property (solubility—coil expansion; insolubility—coil contraction) is illustrated in Figure 6-1.

In a polymer sample (either solution or bulk), random coils interact with other coils and entangle. *Entanglement* is the main reason plastics have mechanical strength. It is also the primary reason why polymer solutions are "thick" (have relatively high viscosities; see "Thick as Molasses" later in this chapter). Even though a polymer might be soluble in a given solvent (thermodynamic statement), it might take a very long time for a sample of it

Figure 6-1. Depiction of a polymer random coil expanded in a solvent or contracted in a nonsolvent.

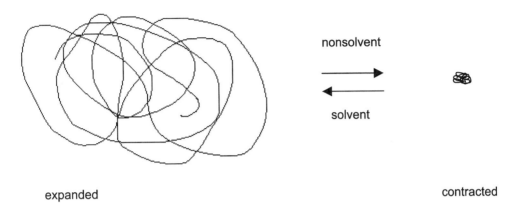

nonsolvent

solvent

expanded   contracted

to actually go into solution. Solvent must first diffuse through the sample, penetrate the chains and swell them, and then decrease the extent to which they entangle with each other. When dissolving a polymer, agitation of the sample almost always speeds the process along.

Why Won't This Stuff Dissolve?

Consider cellulose and starch (starch as we saw in Chapter 3 being a mixture of amylose and amylopectin). Because all of these substances are basically homopolymers of glucose, we would expect them to be soluble in water. Amylopectin is soluble, while amylose *disperses* in water (breaks up to form very small particles) but does not totally dissolve. Cellulose is totally insoluble. Even though all three have the same monomer repeat unit, apparently something in their polymer *structure* functions to control solubility. The molecules of cellulose are linear, and the chains interact with each other to form rigid *crystalline* domains held together strongly through hydrogen bonding. In the case of the sugar dextrose, water molecules easily penetrate the crystals and separate (solvate) the individual glucose molecules. Water is unable, however, to penetrate totally the tightly bound, highly ordered crystalline regions of the polymer cellulose. Therefore the polymer absorbs water but is insoluble in it. Linear amylose is not as highly crystalline as cellulose, but rather forms helices that are held together with hydrogen bonding and that do not totally dissolve in water. Finally, amylopectin, being branched, has more difficulty crystallizing or forming some regular structure. Amylopectin does dissolve in water. Recall that potatoes contain significant amounts of starch. If water that was used to cook potatoes is allowed to sit for a day or so, a semisolid settles out. This is an amylose-rich fraction that separates after a number of helices of amylose come together, crystallize, and precipitate.

Similar effects are seen with synthetic polymers. *Amorphous* polymers are substances without any crystallinity and are often soluble in at least a few solvents. Most crystalline polymers resist dissolving in common solvents. Some can be dissolved, but often with difficulty. If they do dissolve, it is usually in a strong solvent and often after being heated.

Thick as Molasses—Viscosity

So what about the solubility of amorphous polymers? Consider packing "peanuts," those S-shaped polystyrene foam pieces that tend to fly around and seem to stick to everything. Polystyrene dissolves in acetone, an organic solvent found in nail polish remover. If we were to drop one peanut into a liter of acetone, the peanut would seem to disappear. Solvent molecules would first penetrate the foam and surround the polystyrene molecules. This would cause the polymer coils to expand and form a solution in which the polystyrene molecules would be separated from each other. The result would be a very dilute solution. Compare this to the situation in which we added a huge number of peanuts to a liter of acetone. The acetone would break down the peanuts, but the concentration of polystyrene would be much higher than the previous case. Acetone molecules would still surround the polystyrene molecules, but at some critical concentration, the polymer molecules would begin to interact or entangle with each other. As we poured this solution from one beaker into another, we would notice that the solution would flow more slowly and would appear "thicker" than acetone alone. In other words, the solution would have a high *viscosity*. Viscosity is defined as the resistance to flow and is a property of both solutions and pure liquids. Sometimes people refer to it as the stiffness or internal friction of a substance. *Rheology* is the study of the flow or deformation of materials.

In fact, one reason we use polymer solutions is because of their viscosity. Examples include engine oil, paint, glues, shampoo, and hair conditioner. Say you were painting the walls in your bedroom. If the paint had the viscosity of water, it would be very difficult to apply and keep from running right down the wall—or down your arm. If your shampoo or hair conditioner had too little viscosity, it would feel watery and just run out of your hair before it was able to do its job. Many personal care products contain polymers to control viscosity and texture.

One polymer that is often used in these products is poly(ethylene oxide) (PEO), a polymer most often prepared by the ring-opening polymerization of ethylene oxide (see Table 5-3). Alternatively, it is sometimes prepared by a step-growth polymerization of ethylene glycol, in which case it is called poly(ethylene glycol) (PEG). PEO/PEG is a water-soluble polymer that can be synthesized to a molar mass in the millions.

Equation 1

$$\underset{\text{ethylene oxide}}{\underset{\text{H}_2\text{C}-\text{CH}_2}{\overset{\text{O}}{\triangle}}} \longrightarrow \underset{\text{PEO/PEG}}{\left(\text{CH}_2\text{-CH}_2\text{-O}\right)_n} \overset{-\text{H}_2\text{O}}{\longleftarrow} \underset{\text{ethylene glycol}}{\text{HO}-\text{CH}_2\text{-CH}_2-\text{OH}}$$

The viscosity of a solution of a given polymer will be dependent upon the concentration of the polymer in that solution. Viscosity depends also upon molar mass. Higher molar mass means greater chain entanglement and therefore higher viscosity. Thinking about this in reverse, we can use viscosity as a method to determine the molar mass of a polymer sample.

Adding even small amounts of extremely high molar mass PEO to water causes enormous increases in viscosity. Using too much will make solutions so viscous they will look more like a gel than a solution. When wet, PEO also feels slippery. It is sometimes added to hair conditioner to make hair easier to comb. The little white lubricating strip on a safety razor is usually a small sample of PEO. During shaving, polymer from the top surface becomes wet and dissolves, providing lubrication.

Viscosity is also dependent upon temperature, decreasing with increasing temperature. Motor oils need to have a high enough viscosity that they provide a lubricating film between moving parts in an automobile engine. If the viscosity is too low, lubrication is poor. If the viscosity is too high, it will be very difficult starting a cold engine. The Society of Automotive Engineers (SAE) has standardized a series of grades from low (5W) to high (140) viscosity (Booser 1995). The grades for gasoline engines in automobiles usually range from 5W to 40. An oil with just one viscosity (e.g., SAE 10W) might work well in a cool engine but become too "thin" as the engine heats up. Therefore, most commercial oils are "multiviscosity grade" (or multigrade), meaning the viscosity changes less throughout the temperature range to which the oil is subjected. An SAE 10W–40 oil, for example, has the viscosity of a 10W grade oil when cold and the viscosity of a 40-grade oil at normal engine temperatures.

How is this done? Most commonly a low concentration of a linear polymer such as poly(methyl methacrylate) is added to the oil, minimizing the decrease in viscosity as the temperature increases. Another technique utilizes a diblock copolymer in which one of the blocks forms an insoluble dispersion at low temperatures but dissolves at higher temperatures, increasing the viscosity of the solution.

Gels

So far we have seen that amorphous polymers can dissolve in selected solvents, and that, if the concentration is high enough, the solution can have a relatively high viscosity. Also, crystalline polymers tend not to dissolve or to do so with difficulty.

Besides crystallinity, what else can prevent a polymer from dissolving? Consider a polymer made from the salt of a vinyl carboxylic acid, such as sodium acrylate:

Equation 2

$$CH_2\!=\!CH\!-\!CO_2^- \ Na^+ \qquad \longrightarrow \qquad \left(\!CH_2\!-\!\underset{\underset{Na^+}{\overset{|}{CO_2^-}}}{CH}\!\right)_{\!\!n}$$

The monomer is completely soluble in water. We would predict that the homopolymer, too (sodium polyacrylate), would dissolve readily in water, because each repeating unit contains an ionic sodium carboxylate group. (Most sodium salts dissolve in water.) If we were to add, say, one gram to 400 to 600 mL of distilled water and stir, the result would be a gel, not the free-flowing solution that we expected. A *gel* is defined as a chemically or physically crosslinked polymer that is highly swollen by solvent. The solvent is held tightly by the polymer network and does not flow. If the solvent is water it is called a *hydrogel*.

How can we explain such behavior? The sodium polyacrylate acts as if it "wanted" to dissolve in the water, but, because of the interactions of the large numbers of ionic groups with each other, an ionic network sets up instead. The sodium carboxylate acts like an ionic crosslinking agent. By adding a small amount of an ionic salt, such as NaCl, we can break up this network and form a free-flowing solution. Polymers such as sodium polyacrylate, sometimes called *superabsorbent polymers*, find use in applications where absorption of water is necessary, such as in diapers, surgical sponges, and women's hygiene products. In addition, it is incorporated in pads to absorb liquids in packaged meats, added to potting soil to help retain moisture, and placed in cartridges to absorb water from gasoline, diesel and jet fuel.

Gelatin desserts (e.g., Jell-o) are very dilute aqueous mixtures of natural protein from collagen plus flavoring and food color. Globular proteins such as gelatin dissolve in water but interact strongly with each other and "set up" or establish a network of physical crosslinks such as hydrogen bonded helices. These networks can usually be broken up by heating them. Temperature-dependent gels such as this are called *thermoreversible gels*. Gelatin from cow bones, pigskin, and other animal sources has been the dispersing medium for silver halide and other photographically active compounds in photographic emulsions for more than a hundred years.

Polysaccharides from plants, too, can form gels in water. Pectin is used to help gel jams and fruit preserves. Some polysaccharides are used to thicken foods. Alginic acid, extracted from brown algae, is a linear polymer containing many carboxylic acid groups. The sodium salt is soluble in water and is used as a thickener in the preparation of ice cream and other foods. If a sodium alginate solution is mixed with calcium ion, the polysaccharide precipitates as a gel (see Section 4). Agar and carrageenan are polysaccharides from red algae that are also used as thickeners and gelling agents.

What can we do with hydrogels? If we could fabricate a thin piece into a curved disc, we would have a soft contact lens. Originally, contact lenses were made from a rigid, optically clear plastic such as poly(methyl methacrylate) (PMMA). Although they provided good visual correction, they could not be left on the eye for too long without doing damage. PMMA is a good barrier for many gases, including oxygen. Therefore, PMMA ("hard") lenses prevented oxygen in the air from reaching the cornea where it is essential for the metabolism of glucose. Many soft lenses are prepared using a water-soluble monomer, hydroxyethyl methacrylate (HEMA), and a crosslinking agent. These are examples of chemically crosslinked gels. The lens cannot dissolve but absorbs large amounts of water to form a hydrogel. The more water in the lens, the greater the permeability to oxygen, and, significantly, the more fragile the lens. Extended-wear lenses are often thinner versions of daily lenses. Being thinner, they allow more oxygen to pass through the lens, enabling the wearer to keep the lens in for significantly longer periods of time. Some newer contact lenses are prepared from polydimethylsiloxane, which has very high oxygen permeability.

Thus we see that polymers find many useful applications not only as true solutions but also as gels.

Things That Stretch or Flow

Solutions, including many polymer solutions, are well behaved. If poured from a container they *flow*. If they are quite viscous, they will still flow, just more slowly. Solutions such as these are called *Newtonian fluids*. Consider water flowing out of a garden hose. Newton's second law states in essence that the flow of water depends directly upon the pressure behind it. If we double the pressure, we get twice the flow (the water coming out of the end shoots twice as far). The viscosity remains constant and is independent of the pressure (force/unit area). As we mentioned above, the viscosity of many solutions depends upon the nature of the components, their concentrations, the temperature, and, if the solute is a polymer, its molar mass.

When we pour a solution, or stir it or shake it, we are applying a stress to the solution and are deforming it. For a Newtonian fluid, this *deformation is irreversible*. If we pour some olive oil from a bottle into a frying pan, the liquid flows across the bottom of the pan, assuming a new shape. On the other hand, if we stretch a rubber band and then release it, the rubber band returns to its original shape. This deformation is completely *reversible*, and is called *elastic*. These two types of deformation, reversible (elastic) and irreversible (inelastic or viscous), explain the behavior of a wide variety of materials when a force (stress) is placed upon them.

Non-Newtonian Fluids—Weird Behavior

Some materials, including solutions of certain kinds of polymers, however, behave strangely. When subjected to some force, their viscosities can change, and in weird ways. Such materials are said to be *non-Newtonian* materials. Consider two funnels, one containing honey and the other mayonnaise (Figure 6-2). Although both are viscous fluids, only the honey flows from the

Figure 6-2. Honey (a Newtonian fluid) and mayonnaise (a non-Newtonian fluid) in a funnel (courtesy of Dr. R. R. Eley, ICI Paints).

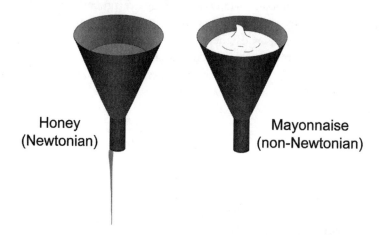

Honey
(Newtonian)

Mayonnaise
(non-Newtonian)

funnel. The mayonnaise will not go through the funnel on its own. Now spread a little of each on a slice of bread. The honey remains very viscous and tends to tear the bread unless we are very careful. The mayonnaise, on the other hand, spreads very easily, doesn't tear the bread, and seems to become less viscous as we work it with the knife. When we stop spreading it, it stays in place. The mayonnaise is a non-Newtonian fluid. We will want to try to understand this unusual behavior.

Let's designate a force such as stirring or spreading a *shear force*. By shearing we mean forcing the molecules to slide past each other. No matter how hard we stir a pot of water or a jar of honey at constant temperature, the viscosity remains the same because both are Newtonian liquids. But what happens when we stir some polymer solutions? It depends. Not all of the examples we discuss below are true solutions. So we'll use the broader term "fluid."

Shear-Thinning and Shear-Thickening Fluids

Some fluids are called *shear thinning*. Common household examples include mayonnaise and latex paint. As they sit in a jar or can, these materials are quite viscous, almost gel-like. However, the harder we stir them, the lower their viscosities become (up to a point). Latex paint sticks to a brush or roller, but flows easily when we apply the brush or roller to a surface. Once

124

it has transferred to a wall or ceiling, it is no longer under the shear force (stress) and the viscosity immediately rises again. That is why paint can spatter and drip when you are applying it but doesn't run down the wall (as water would). Another good example of a shear-thinning liquid is ballpoint pen ink. When you apply the ink to a sheet of paper, it flows nicely. But it does not drip from the tip of the pen, even when you flick it.

Other materials are *shear thickening*. Stirring them or moving them causes the viscosity to actually increase. In some cases, applying a very rapid force will make the fluid seem like a solid. Examples include starch solutions, some printers' inks, wet sand at the beach, and quicksand (Walker 1978). Ever notice when you walk on damp sand that your footsteps appear dry compared to the surrounding sand? Stepping on the fluid (sand and water) applies a force that causes the sand to move. Because the grains of sand are closely packed in the undisturbed fluid, the only direction they can go when you step on them is upward. So stepping *down* causes the sand to move *upward*, creating space below for the water to go. Therefore the sand appears momentarily dry where your foot was.

It gets even weirder. For some materials, the amount of the viscosity increase or decrease depends upon how long the shear force was applied. Once the force is removed, the original viscosity returns, but after some time. Consider catsup, which is shear-thinning. You have probably noticed how difficult it is to pour catsup from a full bottle. (Most restaurants keep their catsup bottles filled!) Turn the bottle upside down, and nothing comes out. Assuming it is in a glass bottle, you need to do one of two things. Pound the bottom of the bottle with the heel of your hand (apply a shear force to the fluid inside), and some will finally flow out. Or stick a knife into the neck of the bottle and scoop some out. If, on the other hand, the bottle is not completely full, you can shake the bottle (again, apply a shear force), the catsup becomes runny (the viscosity decreases) and will flow through the opening. Of course, if the catsup is in a plastic bottle, squeezing the bottle forces it out.

The longer you shear catsup, the runnier it becomes and the longer it remains so. We call such a fluid *thixotropic*. Other examples include margarine, latex paint, and shaving cream.

Viscoelastic Fluids

Finally, some fluids that undergo viscosity changes on shearing are elastic as well. These are termed *viscoelastic* fluids. These materials have properties of both a liquid and a solid. An excellent example is silicone putty (e.g., Silly Putty), which shows three different types of behavior depending upon the shear rate. If a piece of this material is suspended (gravity, a low shear force), it will slowly flow downward like a very viscous fluid. If it is sheared faster, it has rubbery behavior. You can observe this by rolling some of it into a ball and bouncing it on the floor. With very rapid shear, however, the putty becomes brittle and will break. Try pulling a piece of it apart very rapidly. The elastic character disappears at very high shear rates. Pulling it apart slowly causes it to flow and stretch. Pulling it apart twice as fast does not cause it to flow twice as fast. Pulling it very rapidly causes it to act like a solid.

Explanations of this behavior are difficult. Perhaps we can begin to understand viscoelasticity by considering a dilute solution of PEO in water. If we begin to pour this solution from a raised beaker into a beaker placed below it, a stream of viscous fluid will connect the two beakers. On tipping up the raised beaker, so that fluid would ordinarily no longer flow out, the viscous PEO solution will continue to flow *up*, out, and down to the lower beaker. This effect is called a tubeless siphon because the liquid is not enclosed in a hose or tube one would use to siphon an ordinary liquid. Cutting the PEO strand with scissors will cause the upper portion to pull itself back into the top beaker.

The PEO solution is showing elastic properties. As we pour the solution, we are applying a shear force that tends to line up the very long PEO molecules. When this force is removed (e.g., when the strand is cut), the molecules recoil to a lower energy state (random coils), and in so doing cause the fluid above the cut to flow up.

Many non-Newtonian materials such as latex paint and mayonnaise set up temporary structures when at rest. These structures are often the result of intermolecular interactions such as hydrogen bonding. Small, fairly rapid deformations will cause elastic behavior. Slime and mayonnaise "wiggle" if shaken. They deform and then recover. Larger stresses exerted over longer times can break up the structure, causing a decrease in viscosity and irreversible deformation. When the force is removed, intermolecular structure can be reestablished, although this might require some time. All of this complicated behavior depends upon such basic phenomena as the structure of the polymer, the nature and strength of intermolecular interactions, and the rate and magnitude of the force applied.

Some Applications of Polymer Solutions and Dispersions
Films

Polymeric films are used in a variety of applications, including food packaging, shrink-wrap, garbage bags, and photographic film base. Although we will see additional ways of making films in Chapter 8, one technique involves *casting* a film from a polymeric solution. This method is particularly useful for polymers that decompose when heated to melt them. In addition, film casting is important in specialty applications in which very thin, uniform films are required, such as in integrated circuit fabrication (see Chapter 4).

To cast a film, a polymer solution of the appropriate viscosity is spread evenly onto a substrate and the solvent is allowed to evaporate at a controlled rate. The operation can be carried out either by hand or on a coating machine. In manufacturing, film is produced in a continuous process. The coating solution is deposited from a slotted die onto a drum or belt that is moving past the die at a constant speed. After preliminary drying, the film is peeled from the substrate and conveyed into a drying chamber where the remainder of the solvent is removed. See Figure 6-3. The support or backing for 35 mm photographic film is produced by this method. The polymer used, cellulose triacetate, would thermally degrade if subjected to a melt process (see Chapter 8).

Figure 6-3. Schematics of casting a film from solution. a, casting onto a wheel. b, casting onto a belt. (Allcock, Harry, and Fred Lampe. 1990. *Contemporary Polymer Chemistry.* 2nd ed. By permission of Pearson Education, Inc., Upper Saddle River, NJ.)

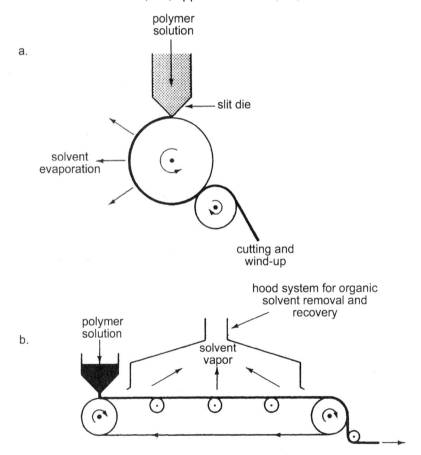

Coatings
Oil-Based Paints and Varnishes
Paint is an old technology that developed around natural materials such as unsaturated vegetable oils (drying oils), mineral pigments, and natural resins. Varnish is paint without any pigment. An example of a drying oil is

linseed oil (from the flaxseed) that has large amounts of long hydrocarbon chains containing one or more carbon-carbon double bonds. When the paint or varnish is applied as a thin coating and exposed to oxygen, the double bonds in the drying oils react by a free radical mechanism forming a crosslinked film. Gradually, synthetic ingredients have replaced natural ones, including a range of synthetic polymers. In a modern oil-based paint, in essence polymeric binder is dissolved in an organic solvent system in which pigment particles are suspended. A number of additives are also present to control properties. As the paint coating dries, obviously, the organic solvent evaporates and enters the atmosphere. Because of the effort to reduce volatile organic compound (VOC) emissions, oil-based paints are being gradually phased out in preference to water-based, or latex, products.

Latexes

In Chapter 5 we discussed emulsion polymerization, the preparation of polymer latexes in an aqueous medium. We noted that this polymerization method had the advantages of using water rather than an organic solvent as the liquid medium, enabling a rapid polymerization to give polymer of controlled molar mass. In addition, the process produces a product in which the polymer is dispersed in water and is ready for use. Isolation and purification steps are not necessary. Let's consider some of the important uses of one class of latex polymers, "acrylics." The largest application for acrylic polymer latexes is for interior and exterior paints. Latex paints for use in homes and businesses are generally prepared with ethyl acrylate, butyl acrylate, methyl methacrylate, and styrene monomers, the mix depending upon specific application.

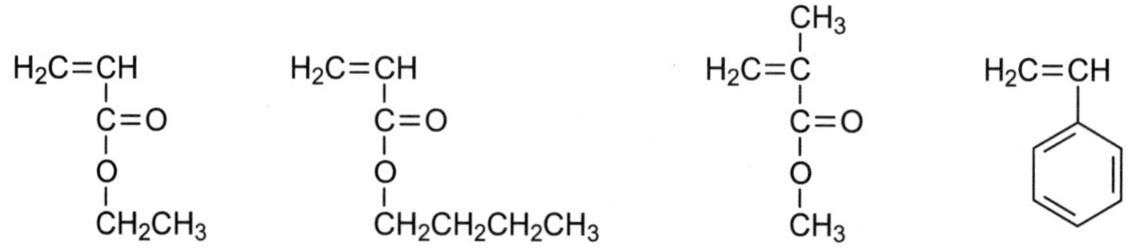

ethyl acrylate   butyl acrylate   methyl methacrylate   styrene

Other monomers might be added to improve adhesion, provide resistance to grease or food stains, improve gloss, or impart other properties. Additional components in latex paints include pigments, viscosity modifiers to optimize application, as well as miscellaneous additives to improve the properties of the final coating. On being applied to a surface as a thin film, the water evaporates, causing the latex particles to come together or *coalesce*. The result is a continuous film in which the pigment is dispersed in the continuous polymer matrix.

The other major uses of acrylic latexes are in textile finishing (e.g., stain and wrinkle resistance), adhesives, and floor polishes and waxes. Adhesives are used in many applications, including pressure-sensitive tapes, contact adhesives, and construction adhesives for carpeting and floor tiles. Peelable labels and stamps probably got their start with the Post-it note, invented by Spencer Silver and Art Fry at 3M company many years ago.

Post-it Notes

For scientists with open, curious minds, failed experiments often lead to new opportunities. The evolution of the Post-it note is an excellent example. In 1968, a chemist at 3M by the name of Spencer Silver was attempting to synthesize new, strong adhesives. One particular sample had very unusual properties. Although the polymer did stick to a variety of materials, it easily peeled off, leaving behind no residue. Fortunately, Silver, kept the "failure" rather than discarding it as waste. At the time, no one could figure out a marketable application for the interesting new material.

A few years later, Art Fry, a 3M new product developer and church choir singer, was having trouble keeping the pieces of paper he used to mark his hymnal pages from falling. During a dull sermon, he began thinking about the falling bookmark problem and remembered Silver's adhesive. Voila! One should be able to coat some of this on a bookmark, temporarily stick it to the desired page, and then later remove it without harming the paper in the hymnal.

Fry made up some samples that he shared with colleagues at 3M. Demand grew, and several new uses evolved. It appeared that a new product was in the making. It took several years, but the self-stick, repositional sheets of paper we now know as Post-it notes eventually became a popular and essential product in every home, school, and office.

Latex gloves, balloons, condoms, and other elastomeric products are produced by dipping metal molds into a latex that can consist of either natural or synthetic rubber. The mold is pretreated with a chemical that causes the latex to coagulate on the mold. Then the latex-covered mold passes through an oven where the elastomer cures (crosslinks). A modern factory can produce on the order of a million latex gloves or balloons a day.

Neil Tillotson and the Latex Balloon

We've seen rubber balloons all of our lives but have probably never thought about how they are made. In the early nineteenth century, toy balloons were made from pig bladders. Later, they were constructed from solidified natural rubber. Today's latex balloons were invented about 70 years ago by Neil Tillotson, a chemical engineer from New England. Tillotson was having difficulty fabricating inner tubes for tires from latex. On a whim, he cut a piece of cardboard into the shape of a cat's head, dipped it into a batch of latex, and let it dry. After removing the dried latex from the cardboard, he had a rubber bag that when inflated, created a balloon with a recognizable shape, including ears. Many people wanted to buy one of his "cat balloons," and a lucrative business was born.

Latex balloons are still made in essentially the same way. A metal form shaped like the uninflated balloon is first coated with a chemical that will cause the latex to coagulate on it. The form is immersed upside-down into a vat of latex also containing a crosslinking (vulcanizing) agent, and then dried in an oven to cure the rubber. After the balloon is removed from the mold, it is ready to be imprinted or packaged. A modern, automated factory can turn out a million or so latex balloons a day.

Tillotson was a Yankee gifted with considerable ingenuity. Born in Vermont in 1898, he joined the U.S. Army as a teenager and served with General John Pershing along the Mexican border. Among other exploits, Pershing's troops (including Tillotson) entered Mexico to capture the bandit Pancho Villa. Tillotson returned to New England to work for a rubber company before starting his own balloon company. He then turned his attention to rubber gloves, founding the Best Manufacturing Company, still in existence. Eventually his varied business interests brought him to Dixville Notch, New Hampshire, where he worked full-time until his death at 102 in 2001. You may have heard of the tiny community of Dixville Notch—its citizens are always the first to vote in presidential elections (beginning at 12:01 AM on election day). In fact, from 1964 through 2000, Neil Tillotson was the very first American to cast a vote for U.S. president.

Summing Up

Some polymers, just like many organic compounds, dissolve in solvents if it is energetically favorable for them to do so. When polymers dissolve, their random coils become greatly expanded because the chains are surrounded by (solvated by) solvent molecules. Crystalline polymers are usually not soluble, except in aggressive solvents and/or if heated.

Polymers that are soluble produce solutions with useful and sometimes unusual properties. The solutions of high molar mass polymers are characterized by having relatively high viscosities. Viscosity is dependent upon molar mass, concentration, and temperature. Polymers are useful in many commercial products because of their relatively high viscosity.

Crosslinked polymers cannot actually dissolve, but some can form swollen gels. Linear polymers can form gels if they have functional groups that can be made to interact strongly with each other. Examples include the protein gelatin, the polysaccharides pectin and alginic acid, and sodium polyacrylate, a polymer sometimes called a superabsorbent polymer. Soft contact lenses are hydrogels (crosslinked polymer swollen with water) produced as thin, curved discs with desirable optical properties and oxygen permeability.

The solutions of some polymers have very unusual behavior. These fall in the category of non-Newtonian fluids, substances whose viscosity varies with the force applied to them. Some fluids become thinner when stirred or shaken, while others become thicker. Still others show solid-like behavior because they have elastic properties under some magnitudes of stress (are viscoelastic).

Polymer solutions and dispersions find many practical applications. We can cast films from some polymers, and we use many others to form coatings. Coatings can be produced from either solutions or dispersions such as polymer emulsions (latexes). Examples include paint, varnish, textile finishes, adhesives, and floor waxes. Other useful latex products are latex gloves, condoms, and balloons.

References Cited

Allcock, H. R., and F. W. Lampe. 1990. *Contemporary polymer chemistry*. 2nd ed., 502. Englewood Cliffs, NJ: Prentice Hall.

Booser, E. R. 1995. Lubrication and lubricants. In *Kirk-Othmer encyclopedia of chemical technology*, ed., J. I. Kroschwitz, 4th ed., vol. 15, 470–88. New York, NY: Wiley Interscience.

Walker, J. 1978. The amateur scientist: Serious fun with Polyox, Silly Putty, slime, and other non-Newtonian fluids. *Scientific American* 239 (5): 186, 188, 191–94, 196.

Other Reading

Coultate, T. P. 1989. *Food: The chemistry of its components*. 2nd ed. London: The Royal Society of Chemistry.

Elias, H-G. 1997. *An introduction to polymer science*. Ch. 6, 7. New York, NY: Wiley-VCH.

Sperling, L. H. 1986. *Introduction to physical polymer science*. Ch. 4. New York, NY: John Wiley.

Stevens, M. P. 1999. *Polymer chemistry: An introduction*. 3rd ed., ch. 2. New York, NY: Oxford University Press.

Chapter 7

Physical Properties

I n several of the previous chapters we have talked about the strength of polymers and pointed out some of the unusual properties that make them both interesting and extremely useful materials. Now it is time to try to understand some of the fundamental reasons for this behavior. We will examine the structure of polymers a bit more carefully so that we can better understand their physical and mechanical properties and discover why plastics such as HDPE (high-density polyethylene), PS (polystyrene), and Kevlar are so different from each other. In this case, the "structure" on which we will focus is the *morphology* of the sample. Morphology refers to the form and shape of a substance (for example, different kinds of rocks have different morphologies). We will see that polymers characteristically have amorphous and/or crystalline morphologies.

The Viscoelastic State

Most of us think of matter as being segregated into one of three states: gas, liquid, or solid. *Gases* flow easily and expand to fill their container. *Liquids* also flow, but they take the shape of their container and have a fixed volume. *Solids* have a fixed shape and do not flow. As a general rule, polymers do not fall neatly into any of these categories. Most uncrosslinked (or lightly crosslinked) polymers do not behave exactly either as liquids or solids. Their behavior has characteristics of both liquid and solid states. We call this state the *viscoelastic state*, the same term we applied to certain solutions in Chapter 6.

A viscoelastic material behaves like either an elastic solid or a viscous liquid, depending upon the force placed upon it. If a force is applied and then quickly removed, the material will deform but then return to its original shape (*elasticity*). If on the other hand, a force is applied and held for some time, the material will deform permanently (*viscous flow*). For example, consider a thin piece of poly(methyl methacrylate) clamped onto a table so that one end extends beyond the edge of the table surface. If we strike the end of the strip quite hard with a hammer, the plastic will break, behaving like a brittle solid. If we lightly hit it with a hammer, the plastic will bend and then snap back, like an elastic solid. Likewise, if we place a weight on the end, the plastic will bend down. If we leave the weight there for a very long time, the plastic strip will remain bent and will not fully return to its original flat shape when the weight is removed. This is the result of viscous flow.

In this chapter we will see that the viscoelastic nature of polymers is responsible for many of their unique properties. In addition, we will see how polymer scientists and engineers have learned how to manipulate polymer structure (chemical and morphological) to optimize some of those properties. So let's begin by understanding something about polymer morphology.

Structure and Thermal Properties
Amorphous polymers and the glass transition

In Chapter 6 we mentioned that the chains of many polymers exist in random coils. In a sample of pure polymer, parts of the chain of one coil penetrate other coils, causing entanglement. Recall that we presented entanglement as a reason for the strength of many polymers.

Amorphous polymers are those whose chains show no order. They exist in entangled random coils. We have used a plate of spaghetti or a bag of Easter basket grass as crude models for amorphous polymers. Note that, based on their diameters, neither the noodles nor the paper or plastic strands are nearly long enough to represent a polymer chain. However, they become physically entangled when mixed together.

For actual polymers, let's consider polystyrene (PS) and poly(dimethyl-siloxane) (PDMS or *silicone*), two common examples. What are some of the differences between our two amorphous polymer examples? Silicone is a clear viscous liquid that "sets" only after it has been crosslinked. (Silicone caulk that is used around windows or bathtubs contains a mixture of ingredients. Basically, however, it flows from the tube as a liquid and then forms a rubbery seal on curing.) PS is a clear brittle plastic that is often used in rigid packaging or inexpensive toys, or is molded into the airline drinking tumblers we discussed in Chapter 1. Why is one a liquid while the other behaves more like a solid at room temperature? If we were to place the silicone in liquid nitrogen (b.p. −196°C), it too would behave as a brittle solid that would shatter if hit with a hammer. If we were to heat the polystyrene above 100°C or so, it would begin to soften and, at higher temperatures, flow like a viscous liquid. In other words, these two very different polymers undergo similar transitions, just at very different temperatures.

In an organic compound, most of the atoms in the molecule can usually rotate with respect to one another, producing what are called different rotational isomers or *conformations*. The rate of these rotations is dependent upon temperature. At room temperature, many of the bonds in organic compounds, small-molecule or polymer, are constantly rotating, producing a variety of conformations. An example is shown below for a rotation about one specific silicon-oxygen bond in a small disiloxane. The R's represent some small organic group such as methyl, CH_3.

Equation 1

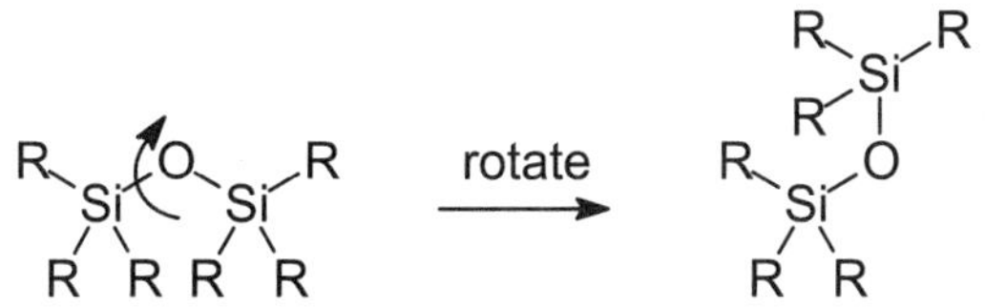

For this rotation, consider holding the left-hand silicon still and rotating up the right-hand portion of the molecule about the Si-O bond as shown with the arrow. (These three-dimensional structures are difficult to illustrate in two dimensions.) While the "product" might look different on paper, because of the symmetry of this simple molecule this rotation produces a structure that is exactly the same as what we started with—two R_3Si groups bonded to a central oxygen atom in a "bent" molecule. Rotations in more complicated compounds can change the shape of the molecule. However, in many small molecules, changes in shape often produce only minor changes in properties.

Now let's see what happens in a polymeric siloxane. Drawn below is a segment of PDMS in which the two CH_3 groups bonded to each silicon have been left off for clarity. They would appear at the end of each of the single lines:

Equation 2

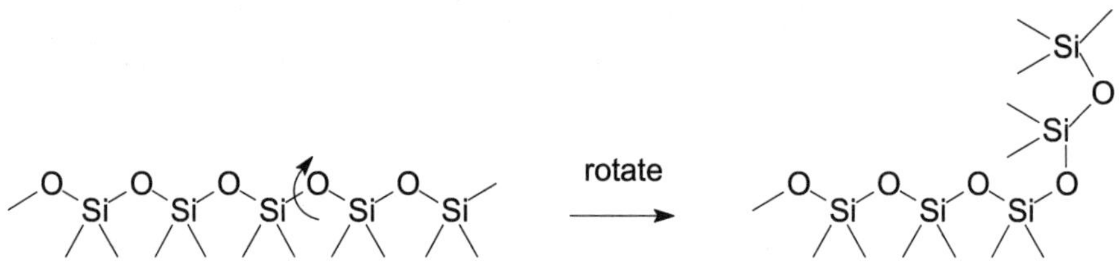

Again, we will hold the left side of the polymer chain still and rotate upward about one Si-O bond. Because this rotation is in the main chain, the shape of the molecule will change as a result of this rotation. If the energy required to cause this rotation is low, many different rotations will occur simultaneously at room temperature and the chains will be quite mobile. The sample will behave like a rubbery material or a viscous liquid. On the other hand, if considerable energy is required to cause rotation about the bonds, the frequency of rotation at room temperature will be low. Examples include polymers with stiff chains or polymers having bulky groups bonded to the chains. These types of polymer chains are not very mobile at room temperature. Samples such as these will have the properties of an amorphous "glass." As the sample is heated, however, at some temperature enough thermal energy will be present to increase the rate of rotation, making the sample more like an elastomer. This is the case with polystyrene, which is a glass at room temperature.

In fact, most polymers have a distinct temperature above which the chains become much more flexible. This temperature can be measured by a variety of techniques and is called the *glass transition temperature*, or T_g. For PDMS it is −127°C, while for PS it is approximately 100°C. It is this tremendous difference in glass transition temperatures that makes PDMS and PS behave so differently (at room temperature). The physical properties of amorphous polymers are determined in large part by the freedom of the chains to move, and this determines the glass transition temperature. Although commercial

136

amorphous rubbery materials have very low T_g's, they do not become liquids above the glass transition because they are extensively crosslinked. A good example is vulcanized styrene-butadiene rubber (SBR; e.g., GR-S) that is often used in rubber bands and tires.

The glass transition temperature has many practical consequences. In the 1950s, one could purchase an inexpensive garden hose that was made from a cheap grade of poly(vinyl chloride). The hoses were not particularly tough, so they tended to tear relatively easily. In addition, however, the T_g of the hose was near 0°C, meaning that people living in temperate climates had to pay attention to the weather. An empty hose left on the driveway would become quite brittle and make interesting crunching noises when the car ran over it after the first heavy frost of the year.

Crystallinity and Melting

In Table 2-1 we saw that for the series of straight-chain hydrocarbons, the boiling and melting points increased as the molar mass increased. Compounds with approximately 16 or more carbon atoms are solid waxes at room temperature and are in fact crystalline. That is, their chains line up and are held together by van der Waals forces in repeating structures. This is depicted symbolically in Figure 7-1a. The forces of attraction exist between the long chains as well as between the chain ends. The melting points of these waxes increase as the chain length increases because the longer chains have more groups that can interact with their neighbors, increasing intermolecular interactions.

Figure 7-1. Depiction of crystalline order for a. hydrocarbon wax; and b. HDPE. Courtesy of Stephen Teegarden.

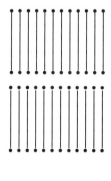

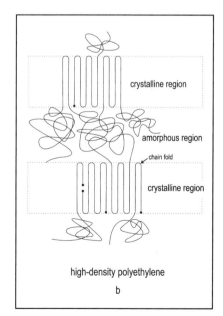

hydrocarbon wax

a

high-density polyethylene

b

We would expect polymers with linear, regular chains such as HDPE also to be crystalline because their chains should tend to pack together in a similar way. However, given their length, it is unlikely that all of the chains can become unentangled and line up as completely as they do in the much shorter waxes. Therefore, "crystalline" polymers generally have domains of crystallinity surrounded by amorphous regions, and are called *semicrystalline*. This is depicted symbolically in Figure 7-1b.

Just as in low molar mass solids, the crystalline regions of polymers are held together by van der Waals forces and exhibit a *melting point*, T_m. At the melting temperature, the crystalline regions, called *crystallites*, are thermally disrupted and undergo disordering into an amorphous liquid. Because most crystalline polymers also have amorphous regions, they have a T_g also, which is lower than T_m. However, polymers generally do not exhibit sharp melting points as small molecules do, but rather soften and melt over a range of 10 to 20 Celsius degrees.

Consider the effect on physical properties this combination of amorphous behavior and crystallinity has. Birthday candles (hydrocarbon wax) are relatively brittle and easily broken. The spout on a laboratory wash bottle, about the same diameter as the candle, is usually fabricated of polyethylene. The spout is tough, meaning it bends but does not break easily (Sperling 1986). The chains in the candle are held together only by weak van der Waals forces, while the crystalline regions of polyethylene are strengthened by the connection of the chains through the amorphous regions. The regions of crystallinity provide a large number of little anchor points in the polymer surrounded by amorphous regions. As a polymer sample is stretched or bent, the rubbery amorphous regions provide flexibility. The amorphous chain segments can be extended only so far, however, because they are connected to crystalline regions. For this behavior to work effectively, the polymer must be used at a temperature near or above T_g but below T_m. Polyethylene has a T_g of approximately $-125°C$ and a T_m of about 138°C. Notice that its T_g is almost the same as the T_g of PDMS, which was one of the examples we used in discussing amorphous polymers. So, although both PDMS and polyethylene have very similar T_g's, one is a liquid while the other is a tough plastic. The significance of crystallinity on mechanical properties cannot be overemphasized.

What do we need to promote crystallinity in a polymer? First, provide linear chains that can align or pack together as we said. These will form crystalline regions based on van der Waals forces. Second, we can strengthen the attractive forces by adding functional groups that can interact with each other, for example through dipole-dipole interactions, through hydrogen bonding, or through ionic interactions. Consider the structures that follow. The first represents the alignment of four linear polyethylene segments in an idealized two-dimensional zigzag conformation. For clarity, the CH_2 groups are not shown but are understood to be located at each break in the line.

Packing of polyethylene chains

The second depicts segments of the polyamide nylon-6. Because they are linear, the chains can align and interact through van der Waals forces as does polyethylene. In addition, each nitrogen has a hydrogen atom that can form a hydrogen bond with the carbonyl group on a neighboring chain. The hydrogen bonds, which are shown by dashed lines in the structure, can form at periodic intervals along each chain. Again, CH_2 groups are not shown.

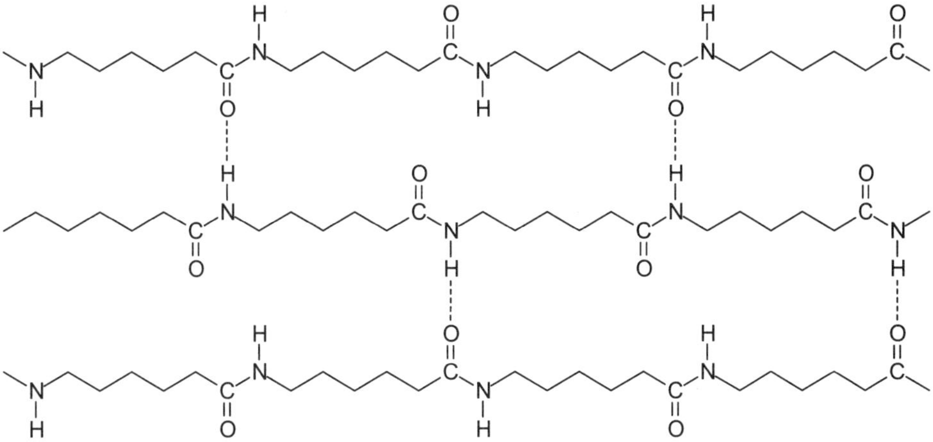

nylon-6

Other types of attractive forces that can help promote crystallinity are dipole-dipole and ionic interactions. The former occur with molecules containing atoms with different electronegativities (for example, a polyester), while the latter are found in polymers containing functional groups that can ionize, such as carboxylic acids. An example of each type is drawn here:

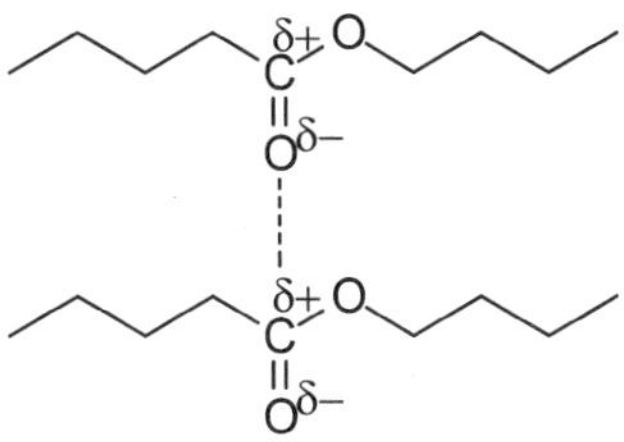

dipole-dipole interaction

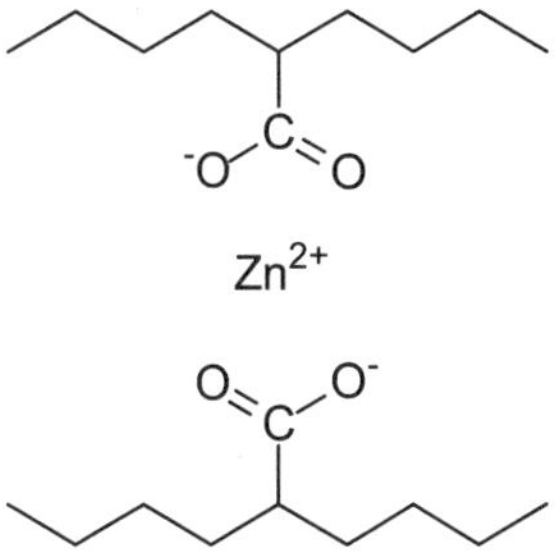

ionic interaction

The dipole-dipole interaction is shown with a polyester but is a factor in many other polymers including poly(vinyl chloride), polycarbonates, and polyethers.

Density

Compounds have unique densities that depend upon their chemical composition, their physical state, and for crystalline solids, the nature of the crystalline lattice. For amorphous polymers, the density of a sample is heavily influenced by the molecular formula. For example, a polymer containing carbon, hydrogen, and halogen atoms has a greater density than a similar polymer containing only carbon and hydrogen (a hydrocarbon) because halogen atoms have higher atomic mass than carbon or hydrogen. Similarly, those containing oxygen have greater densities than hydrocarbons. A few examples are shown in Table 7-1.

Table 7-1. The effect of molecular formula on the densities of a series of amorphous vinyl polymers.

Amorphous Polymer	Repeat Unit Formula	Approximate Density (g/cm^3)
polytetrafluoroethylene	C_2F_4	2.0
poly(vinylidene chloride)	$C_2H_2Cl_2$	1.8
poly(vinyl chloride)	C_2H_3Cl	1.4
poly(vinyl alcohol)	C_2H_4O	1.3
poly(methyl methacrylate)	$C_5H_8O_2$	1.2
polypropylene	C_3H_6	0.85

For semicrystalline polymers the situation is a bit more complicated. We have used the term density in differentiating two different types of polyethylene—low density (LDPE) and high density (HDPE). Now that we appreciate some of the basic differences between amorphous and semicrystalline polymers, it is pretty straightforward to understand relative densities of bulk polymers. For a given polymer composition, the same phenomenon that gives rise to crystallinity (chain alignment) also controls density. The more compact the chains, the greater the mass per unit volume and the greater the density. The chains in amorphous polymers, or amorphous regions in semicrystalline polymers, have more space between them and thus are lower in density. The space between polymer chains is called the *free volume*. LDPE is an excellent example because its chains are branched, reducing the opportunity for crystallinity. Its bulk density is approximately 0.92 g/cm³. HDPE is considerably more crystalline and has a density of approximately 0.96 g/cm³. In fact, one can use the experimental density value to help identify an unknown polymer (see Section 4). Commercial polyethylene is available in a wide range of grades, most of them differing in the amount of chain branching. Because branching and crystallinity are inversely related, one can estimate the degree of branching by measuring the density.

Perhaps it would be best to complete this section with a brief summary. Figure 7-2 generalizes the physical properties of both amorphous and semicrystalline polymers as a function of temperature. Below the glass transition temperature, polymers are in a glassy state. For semicrystalline polymers, the amorphous, glassy regions are mixed with crystalline domains (crystallites). On being heated above T_g, amorphous polymers become rubbery, then gummy, and finally liquid. Semicrystalline polymers above T_g still contain crystalline domains and therefore are flexible and often tough thermoplastics. Above the melting point, T_m, they become amorphous liquids just like their fully amorphous cousins.

Figure 7-2. Physical properties of amorphous and crystalline polymers as a function of temperature. (Allcock, Harry, and Fred Lampe. 1990. *Contemporary Polymer Chemistry*. 2ⁿᵈ ed. By permission of Pearson Education, Inc., Upper Saddle River, NJ.)

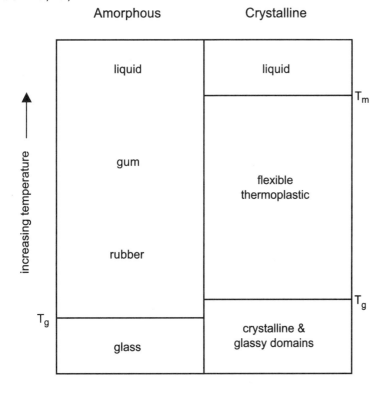

Mechanical Properties
Stress-Strain Properties

We have often used the term *strength* to describe a polymer. But what does this mean, exactly? A rubber band, a plastic grocery bag, and a poly(methyl methacrylate) storm door window are all strong, but in completely different ways. Let's consider pulling on a sample and measuring the energy required to stretch it. We call the energy involved the *stress*, and the distance of stretching the *elongation* or *strain*. Figure 7-3 shows a typical stress-strain plot for three different kinds of polymers. Here we see that a brittle plastic elongates

Figure 7-3. Stress-strain curves for different polymers.

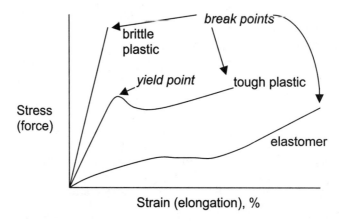

very little (usually only 1% to 2%) as force is applied to it. Eventually the sample breaks (break point). The plot of this data results in a reasonably straight line. The slope of this line is called the *modulus*:

Equation 3

$$\text{modulus} = \frac{\text{stress}}{\text{strain}}$$

Specifically, this experiment uses *tensile* stress, and the modulus is called *tensile modulus*. We need to make this distinction because other kinds of stress are possible (e.g., recall *shear stress* in Chapter 6), as are other types of modulus.

This experiment provides fundamental information on the *stiffness* or *tensile strength* of a material. (Tensile is defined as the capability of being stretched or extended.) Consider by contrast the bottom curve in Figure 7-3, which describes the behavior of a rubbery material. Minimal stress results in substantial elongation, signifying a low tensile modulus (linear portion of curve). Elastomers commonly can be stretched several hundred percent before they break. In the middle is the plot of a tough plastic, a material that combines properties of both of the other two classes. Tough plastics often have a modulus somewhat lower than that of a brittle material but substantially higher than that of an elastomer. At some point, the sample *yields*, meaning it begins to elongate at nearly constant stress. In this region it behaves more like an elastomer rather than as a plastic although the deformation is permanent. Eventually it breaks. This behavior is consistent with that of a semicrystalline polymer above its T_g, such as HDPE.

Toughness and Impact Strength

Often, however, we are not too interested in actually stretching a polymeric material. An engineer might ask questions such as these. In designing plastic lenses for the exterior lights on our automobiles, which optically clear plastics are the least likely to crack when struck by stones thrown up by other cars? Which plastic will best resist breakage when used as a bumper in an automobile? Which elastomer will provide adequate protection against sharp objects and have the desired resilience when used in a tire?

Many applications require a *tough* polymer. What do we really mean by tough? It is usually defined as the area under the stress-strain curve. Therefore the tensile test can provide one measure of toughness. However, for many applications the tensile test, in which a sample is slowly stretched, may not best predict how a material behaves when rapidly stressed. In addition to the tensile test, there are a number of other measurements that provide information about the strength of polymers under varying experimental conditions. One example is *impact strength* or *impact resistance*, the ability of a material to withstand a sudden force without breaking. Figure 7-4 shows an Izod pendulum device for measuring impact strength. A sample of standard shape is clamped in a vise at the base of the instrument. Often a small notch is cut into the sample to provide a standard weak spot at which failure will occur. The weighted arm is released from a predetermined height and swings down, striking and breaking the sample. The smaller the arc through which the arm moves after breaking the sample, the greater the impact strength (the greater the toughness).

Polystyrene has low impact strength, approximately 15 J m^{-1}, while polycarbonate, an engineering plastic, has an impact strength of about 800 J m^{-1}. Quantitative information such as this helps guide the engineer in matching materials with applications. For example, both polystyrene and polycarbonate are optically clear polymers that might be considered for use as an automobile taillight lens. Given the impact strength data, however, the engineer knows that polycarbonate would withstand being struck by stones adequately while polystyrene would prove extremely unsatisfactory. In fact, the impact strength of polycarbonate is so high it is sometimes used for bulletproof glass and for motorcycle, sports, and construction helmets. Recall that polycarbonate is the polymer used for CDs and DVDs (see sidebar on compact discs in Chapter 4).

Figure 7-4. Drawing of Izod pendulum impact device. Close-up: test sample clamped in holder (Yee 1987; reprinted with permission of John Wiley and Sons and the American Society for Testing Materials. Copyright ASTM International.)

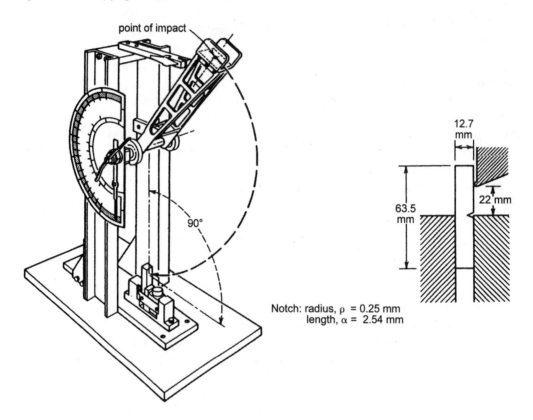

Notch: radius, ρ = 0.25 mm
length, α = 2.54 mm

Improving Polymer Strength
Orientation

We have proposed that crystallinity occurs when segments of linear polymer chains align and interact favorably with each other. When a "crystalline" polymer is cooled from the melt, some of the chains line up and form crystallites because it is thermodynamically favorable for them to do so. The extent of crystallinity depends upon a number of factors, including the polymer structure and the rate of cooling.

Sometimes we can actually encourage chains to align by physically manipulating a polymer sample. You can demonstrate this yourself using narrow strips of film cut from an LDPE dry-cleaning bag or a sandwich bag (see Section 4 for details). If we hold on to each end of a strip and pull quickly, the sample will break. Pulling slowly on the strip, however, will begin to stretch the film. If we release, the film will return to its original length. If we continue pulling, the film will begin to *yield* (see Figure 7-3). At this point, less energy is required to elongate the film, and the strip is made permanently longer. To accommodate the stress placed upon the sample, the amorphous polymer coils disentangle and straighten out (change conformations), becoming aligned in the process. Stretched areas of the sample may become hazy, a visual indication of the formation of crystalline regions. As the strip of film is stretched, it becomes narrower as well as thinner, although the latter may be difficult to see. Further stretching will cause the film to break.

The exercise above is a good example of viscoelastic behavior. In effect, we are observing how rapidly the polymer molecules are able to move. When the strip is pulled rapidly, the chains cannot disentangle rapidly enough and the sample breaks. Slower elongation allows the polymer molecules time to reorient. If orientation is done under controlled conditions (as on a machine at a uniform temperature), a polymer sample (e.g., a film, bottle, or fiber) becomes many times stronger even though the thickness decreases. The process is carried out so that a predetermined degree of crystallinity develops in the sample, a value known to provide maximum strength. We will see specific examples of this in Chapter 8.

As in many areas of our experience, there can be too much of a good thing. In Chapter 4 we discovered that when HDPE was still quite new, pipes and bottles fabricated of the polymer weakened, developed cracks, and eventually failed. It was just at this time that the demand for Hula Hoops skyrocketed, allowing manufacturers time to determine the cause of the disaster. The problem was really one of too high a crystallinity. The polymer made by the Ziegler or Phillips process produced polymer that was too perfect. The solution was the incorporation of a very small amount of a comonomer with a short side chain that breaks up the crystallinity slightly, reducing the cracking problem while maintaining overall strength.

Making Really Strong Polymers

We have seen that different kinds of polymers differ in their basic strengths. Some are inherently weak and brittle, some are very rubbery, while still others are much stronger (e.g., have greater toughness or greater impact strength). It would follow, then, that the polymer chemist should be able to determine the strength of a polymer by his or her choice of monomers used in making the polymer. We have just completed a section in which we have learned some of the phenomena that impart strength. Partial crystallinity is good, as is some chain flexibility. Hydrogen bonding increases interchain attractive forces. Other factors include thermal stability, or the ability of a polymer to maintain its strength at high temperatures. We call this property *heat resistance*. Polymers with many benzene rings in the backbone (aromatic groups) generally have higher heat resistance than those without (having aliphatic groups). In addition, many step-growth polymers are more thermally stable than most chain-growth polymers.

How might we put all of this information together? Consider constructing a step-growth molecule that is highly ordered, contains mostly aromatic groups, and is capable of hydrogen bonding. The synthetic polymer chemist will think of polymers such as polyamides, because they undergo hydrogen bonding very efficiently. Start with a structure such as nylon-6,6, but replace the aliphatic groups with aromatic rings:

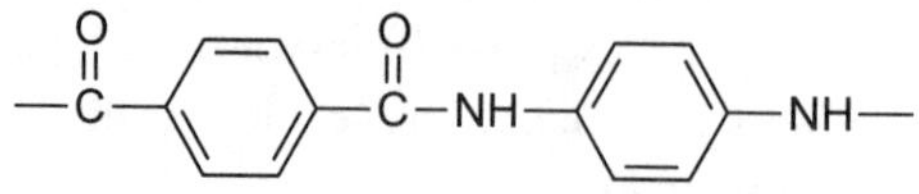

instead of

The result is an all-*aromatic* poly*amide* called an *aramid*. The specific example shown is called poly(p-phenylene terephthalamide), better known by its trademarked name of Kevlar. It has a very high melting point because of its strong intermolecular forces. Figure 7-5 shows the structure of this polymer, suggesting a very regular order with hydrogen bonding between every carbon-oxygen double bond and a hydrogen atom on a nitrogen atom on an adjacent chain.

Figure 7-5. The structure of Kevlar polyaramid. Polymer chains run vertically and are associated with other chains through hydrogen bonding (dashed lines).

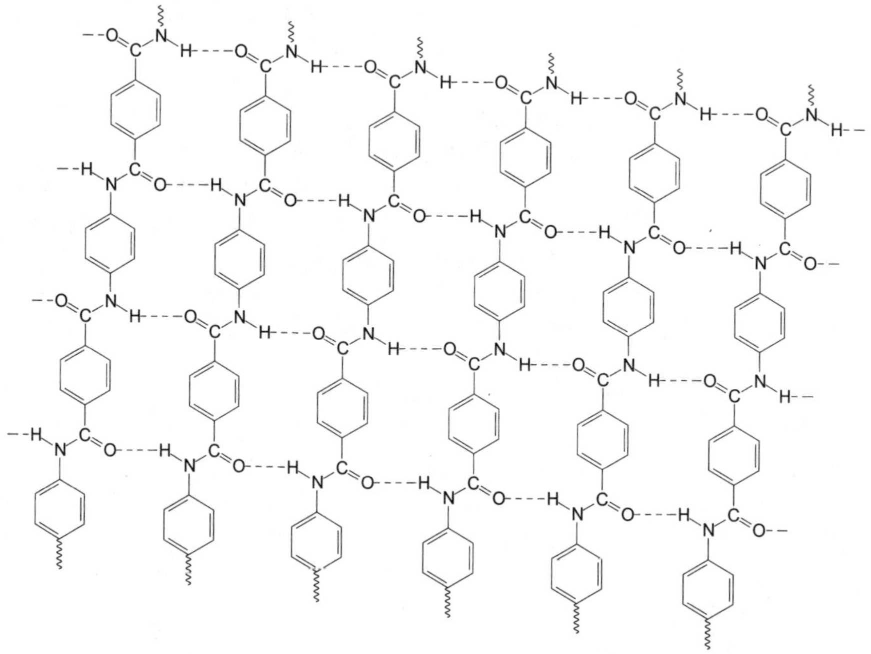

Kevlar is an example of a class of polymers called *high performance polymers*, characterized by their very high heat resistance, high impact resistance, superstrong fibers, and very high cost. They are usually difficult to work with, are synthesized by interfacial polymerizations, and are processed in very strong solvent systems. Kevlar solutions are actually liquid crystalline, a topic we take up in Chapter 10. OK, we can understand some of the approaches to making extremely strong polymers, but they tend to be expensive. Can we get something for almost nothing? Can we improve the strength of "ordinary" polymers?

Tricks for Strengthening Ordinary Polymers

Let's say we have a pretty useful polymer that would be even more useful if it had greater toughness. The first thing we might consider doing is making the polymer chain more regular so that chain alignment and crystallinity increase. Recall that atactic polypropylene is a soft polymer with inferior properties. (See the section "Polymer Stereochemistry: Tacticity" in Chapter 5 for structures of atactic and isotactic polypropylene.) When Natta measured the physical properties of his isotactic polypropylene, he realized its toughness resulted from crystallinity, and that resulted from the stereoregularity along the chain. Structure dictates properties, a statement that is as true for polymers as it is for any compound. Thus we can see why there was so much interest 50 years ago in the new catalysts such as the Ziegler-Natta systems. Interest in new chemistries continues to this day, because of the desire to gain ever more sophisticated control over polymer structure.

If you are a yogurt fan, you no doubt have your favorite brand and flavors. Just as not all yogurts are the same, not all yogurt containers are the same either. The next time you are in the grocery store, look at the tops on the Dannon yogurt packages and guess of what polymer they are made. They are optically clear, tough, and very thin. It might surprise you to find out that the polymer used in this application is actually polystyrene. So far we have often used the words *polystyrene* and *brittle* in close proximity. Can the identification of the polymer used to make the yogurt lid be correct? Just as coordination catalysts can be used to make isotactic polypropylene, so too can they make isotactic polystyrene (i-PS). Natta, in the 1950s, was one of the first to do this. i-PS is a crystalline polymer with a high degree of order. It can be fabricated into sheets, stretched to orient the chains, and made into an inexpensive food packaging polymer.

Recapping, the physical properties of three out of the four cheapest, highest volume commodity polymers can be very significantly enhanced by figuring out how to make their structures more regular: LDPE versus HDPE or LLDPE; atactic-PP versus isotactic PP; and atactic PS versus i-PS. This point underscores the importance of developing synthetic reactions that provide very high degrees of order during polymerization.

Thermoplastic Elastomers

Elastomers represent special challenges because their useful temperature range is always above their glass transition. We have noted they are usually crosslinked so that they return to their original shape after being stretched, rather than simply deforming (flowing) to some new shape. Crosslinked polymers, as we have seen, are thermoset materials and do not melt. Therefore they can be difficult to process. Most elastomers are crosslinked (vulcanized) while being molded. As we will see in Chapter 9, thermoset materials cannot be readily reprocessed and are difficult to recycle.

A clever way to solve these dilemmas is the following. Synthesize an A-B-A triblock copolymer (see Chapter 5) in which the B block is an elastomer (say, polybutadiene) and the A blocks on either side are a crystalline or higher T_g material (say, polystyrene). In Chapter 5 we pointed out that mixtures of two or more polymers (polymer blends) most often consist of separate phases. Even *parts* of chains (e.g., in block copolymers) exclude chain segments of other polymers. In the example above, the butadiene chain segments will associate with each other, while the styrene segments will be excluded by the butadiene regions and will associate only with other styrene segments. The general idea is depicted in Figure 7-6. The polybutadiene blocks are the ma-

Figure 7-6. Simplified morphology of a thermoplastic elastomer. A-B-A triblock copolymer has amorphous elastomeric B blocks (solid curves) between high T_g or crystalline outer blocks (jagged lines). Circles represent crystallites or domains of high T_g outer blocks.

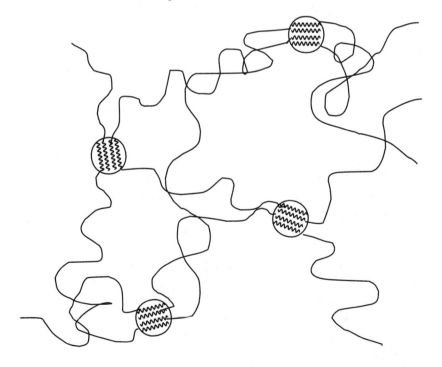

jor component of the sample, which functions as an elastomer at room temperature, because the T_g is about $-100°C$. The polystyrene segments (T_g about $100°C$) are hard, glassy "anchors" that prevent the butadiene from stretching too far. The polystyrene phases act as crosslinks, but the effect is lost if the sample is heated above $100°C$, at which point the polystyrene flows also. This, then, is an example of *physical (temporary) crosslinking* as opposed to the chemical (covalent) crosslinking of elastomers brought about with something like sulfur. These materials can be melted and processed much like conventional thermoplastics. Upon cooling, the "hard" and "soft" phases separate, reinstating the physical crosslinks and locking in the structure. Thus the name *thermoplastic elastomer*. These materials are prepared from a wide variety of monomers and find numerous applications. Some are used in sporting equipment, for example, the soles of sneakers and running shoes. Some polyurethane compositions are processed into elastic fibers such as spandex (e.g., Lycra).

Spandex—Not So Firm a Foundation

We all use *spandex*, that extremely stretchable fiber known as *elastane* in Europe or called Lycra if manufactured by DuPont. It provides the "stretch" in panty hose and other undergarments, swimsuits, jeans, cycling shorts, tights, and many other articles of clothing.

Spandex was developed at DuPont by Joseph Shivers in 1959 as a replacement for relatively heavy rubber threads. Lycra was the computer-generated name chosen for this new product. Spandex fibers, unlike rubber, do not break down in the presence of body oils, perspiration, lotions, or detergents. The first products utilizing this new fiber were much-improved women's panty hose, girdles, and other "foundation garments." Soon thereafter, spandex swimwear appeared, followed by stretch ski suits introduced at the 1968 Olympics. Other sports soon adopted the fiber, causing a revolution in sports fashions. The pop singer Madonna began performing in spandex outfits in the 1980s, initiating a trend in stretch streetwear fashions.

Chemically, spandex is a thermoplastic elastomer. It is a polyurethane block copolymer consisting of long, soft, rubbery segments alternating with rigid, high-Tg segments (see structure below).

soft segment   rigid segment

The soft, low-Tg segments are quite long (x is a large number), with the chains preferring to exist in random coils. The rigid segments, substantially below their glass transition temperature, "anchor" the soft segments and prevent permanent flow or deformation. The result is a fiber that stretches some 600 times its original length and then returns to its original shape.

Stretchable garments usually contain no more than 20% to 25% spandex. The elastic fiber can be blended with many other fibers, including cotton, wool, silk, or synthetics to produce the final fabric. Spandex continues to find new uses, including door panels for automobiles, upholstered furniture, and more comfortable leather shoes.

Additives

Polymer products often contain one or perhaps several *additives*, substances that are incorporated into the sample to alter the polymer's properties or to aid in processing it. These materials function in a number of ways, such as improving mechanical properties, modifying chemical behavior (e.g., stabilizing an elastomer against air oxidation or reducing flammability), adding color, or preventing the buildup of static charge. Stevens provides an extensive list of additives broken down by function (Stevens 1993). We will examine a few of them in this chapter.

Plasticizers

In Chapter 6 we talked about the ability of solvent molecules to interact with and surround amorphous polymer chains, leading to the formation of polymer solutions. A closely related phenomenon utilizes a low-molar mass compound to penetrate a polymer and reduce the forces of attraction between chains. Such a compound is called a *plasticizer*. It must be compatible with the polymer and is almost always nonvolatile. Solvent molecules actually plasticize a polymer sample before forming a solution. However, most solvents are not good permanent plasticizers because they diffuse to the surface and evaporate.

Because it reduces interactions between chains, a plasticizer lowers the T_g of a polymer. Poly(vinyl chloride) (PVC) is a rigid polymer with a T_g of 85°C and a low degree of crystallinity. Products prepared of PVC are almost always plasticized, the amount of the plasticizer, in addition to other additives, determining the ultimate properties. Rigid PVC plumbing pipe or electrical conduit contains minimal plasticizer. At the other end of the spectrum is Tygon laboratory tubing, which contains enough plasticizer to keep the PVC flexible at temperatures far below room temperature. In between is a wide range of products as diverse as vinyl windows, siding, gutters, molding and trim, insulation for electrical wires, wall coverings, tablecloths, shower curtains, toys, food packaging, and others. As we mentioned, garden hoses are also commonly prepared from PVC (and are of much higher quality than those originally introduced to the market in the 1950s).

Historically, plasticizers played a crucial role in the development of the first plastic—celluloid. In Chapter 4 we saw that Hyatt, Parkes, and Spill formulated cellulose nitrate with camphor to produce a solid that could be fabricated into hard, useful, and attractive objects. Camphor is a relatively low molar mass natural product that acts as a plasticizer for cellulose nitrate, reducing stiffness and brittleness.

You are no doubt familiar with the "plastic" odor of certain new items such as vinyl shower curtains. Much of what you are detecting is plasticizer that has evaporated from the surface of the polymer. Recently, attention has been focused on the migration of plasticizers from the plastics that contain them into food and the environment. As a result, *polymeric plasticizers* are often used for applications in which plasticizer migration could be a concern, such as food and medical packaging. Being macromolecules, these compounds are entangled and are unable to undergo significant migration.

Fillers and Reinforcing Agents; Composites

We first encountered fillers in Chapter 1. *Fillers* (also called *extenders*) are used to reduce cost and sometimes also to improve the strength of a polymer. In the latter case they are called *reinforcing agents*. Polymeric materials that contain fillers are called *composites*. Actually the term composite is used more broadly to describe any complex material made up of two or more structurally different components, the properties of the composite differing from those of either individual component. Teeth and bones are examples of natural composites. Humans have probably always tried to make materials stronger, lighter, and cheaper by making *composites*. Natural materials such as sand, rock, and straw have been used to reinforce mud and clay building materials for many thousands of years. Ancient Egyptian, Chinese, and Japanese cultures used straw and horsehair to reinforce mud bricks (adobe). Concrete is an excellent example of a more modern composite, being made up of cement, stones, and/or sand. It is often reinforced with steel rods.

A polymeric composite you are no doubt familiar with is peanut brittle. Corn syrup (polysaccharide) and sucrose are cooked together, producing larger polysaccharides and eventually forming a plastic mass. Peanuts are added, as is baking soda (sodium bicarbonate) to provide small bubbles of CO_2 that decrease the density. A little butter is added, partly for flavor, partly for texture, and partly to reduce sticking. The hot, molten composite is spread out in thin sheets and cooled. In some recipes, the still hot, taffy-like material is stretched to produce a thinner, transparent final product.

A large number of inorganic and organic substances are used as fillers in polymer composites. Calcium carbonate, barium sulfate, clays, silica, and talc are common examples. Glass beads are often used in traffic paints to increase reflectivity. Metal fibers are sometimes added to impart conductivity or to improve metal plating. A number of organic materials are also used, including wood flour, cellulose, and even corncobs. We will encounter starch/polyethylene composites in Chapter 9 when we discuss the disposal and degradation of polymers.

Have you ever seen an automobile tire that was *not* black? (Some have white sidewalls or white lettering but otherwise are black.) Windshield wiper blades and the rubber seals around car doors are also black. The reason is that they contain finely divided carbon black, an excellent reinforcing agent for rubber. Approximately 30% of the mass of a typical tire is carbon black. The carbon particles also act as a pigment. The small particles form physical and chemical bonds to the polymer chains, essentially preventing them from stretching too far. As a result, tensile strength and modulus increase, as does toughness. As the rubber in the tire or wiper blade ages, oxidation begins to degrade the polymer molecules and free carbon black accumulates on the surface. Wiping a rag or a finger across it produces a black smudge.

Fibers

For polymers, glass fibers have been used to make polymer composites for a very long time. Glass fiber (e.g., Fiberglas) is a common and relatively inexpensive material for fabricating boat hulls, automobile bodies, snowmobiles, decorative motorcycle parts, and a large number of household items. Glass fiber mesh is first positioned on a mold and a liquid polyester or epoxy resin is applied over and through the resin and then extensively crosslinked. The resulting object is strong, although it is not very resilient and will break if struck too hard. Chevrolet Corvettes have had glass fiber-reinforced body panels since they first came out in 1953.

Longer, stronger fibers can also be used. Polymer composites containing graphite ("carbon") fibers or aramid fibers, sometimes called *advanced composites*, are relatively lightweight and are often used in applications where very high strength is required, such as in aerospace and military applications. The fibers have extremely high tensile strength. In this case, it is the fiber that provides the strength of the material, with the surrounding polymeric material (the *matrix*) acting as the binder that holds everything together. For these composites, the matrix polymer is usually a thermoset such as epoxy or crosslinkable polyester. Properly fabricated, these composites have mechanical properties equal to or greater than those of most metals and at reduced mass. Commercial airplanes contain large amounts of advanced composites, enabling important mass and fuel savings. The density of aluminum is approximately 2.7 g/cm^3, while that for a Kevlar/epoxy composite is only about 1.4 g/cm^3. Space shuttles use Kevlar composites in the casings for rocket motors. Although relatively expensive, these fibers have also found application in sporting equipment (canoes, kayaks, downhill skis, snowboards, bobsleds, hockey equipment, pole-vaulting poles, and golf club shafts) (Jacoby 2002).

Impact Modifiers

An engineer is looking for an inexpensive polymer to use in an automobile part. He or she knows that polystyrene would be much too brittle. Polycarbonate would be a much better choice, although it costs significantly more. We need a good way to strengthen a polymer like polystyrene inexpensively. We have just discussed several methods for improving the mechanical properties of polymers. In addition to these techniques, one could think about synthesizing copolymers of styrene and less brittle monomer(s). Actually, we have already seen that this approach has been used with considerable success (see Chapter 5 and Table 5-2). Styrene-acrylonitrile (SAN) copolymers and acrylonitrile-butadiene-styrene (ABS) terpolymers have excellent impact strength. Although sometimes copolymerization is a viable option, oftentimes a completely different approach is called for. Let's see how.

When polystyrene is sufficiently stressed, cracks develop that propagate rapidly through the sample, causing failure. Tiny particles of low-T_g elastomers can be mixed into the sample, providing energy-absorbing sites to stop crack propagation. As a tiny crack begins to grow, it runs into a rubbery site, which absorbs the energy and prevents growth into a large crack. These substances, called *impact modifiers*, increase the toughness of the brittle polymer substantially. For example, mixing 5% to 20% butadiene rubber microspheres into a polystyrene sample can raise the impact strength substantially and increase the elongation from about 2% to at least 15%.

Crosslinking or Vulcanizing Agents

As we have pointed out several times, elastomers are used at temperatures considerably above their glass transition temperatures and need to be crosslinked to keep them from permanently deforming. Incorporating physical crosslinks, such as in thermoplastic elastomers, is one approach. Most elastomers are crosslinked chemically, however, using a material that can form covalent bonds between two or more chains. We have mentioned the use of sulfur as a vulcanizing (crosslinking) agent for natural and synthetic rubber (Chapter 4). Although the chemistry of vulcanization is complex and we do not yet completely understand it, its effects on the properties of rubber are clear. Vulcanization transforms the weak elastomer into an elastic network with greatly increased tensile strength, stiffness, and toughness. The tensile strength of unvulcanized (raw) natural rubber is approximately 2 x 10^6 Pa (300 psi) while that for a typical vulcanized material is at least 10 times that. The field of polymers has developed into a sophisticated scientific discipline in terms of the detailed understanding of polymer structures, the ability to synthesize complicated macromolecules, and the development of techniques for fabricating useful materials. It is a great irony that, despite a rather sophisticated understanding of conventional crosslinking chemistry, the original technology discovered accidentally by Goodyear in 1839 is still the predominant method used today for vulcanizing rubber.

Vulcanization is a term for the crosslinking of an elastomer. Other terms for specific applications include *curing* (e.g., crosslinking siloxane caulking materials; forming boat hulls with polyester and glass fiber) and *drying* (e.g., crosslinking coatings such as paints and varnishes).

A large number of crosslinking agents are known and used today. Some react with the double bonds that exist in many elastomers based on diene monomers (natural rubber, polybutadiene, polyisoprene, etc.) Others react with a wide range of polymers through free radical or oxidation reactions. Crosslinking agents are often added during processing, although not always. Adding a small amount of comonomer with two or more reactive groups at the time of polymerization produces a crosslinked polymer network.

Summing Up

The physical behavior of many polymers lies somewhere between that of a liquid and that of a solid. It is called the viscoelastic state. In this chapter we have divided polymers (yet again!) into two classes based on morphology: amorphous and semicrystalline. Two properties for polymers that determine their usefulness are the glass transition temperature (T_g) and, for semicrystalline materials, the melting temperature (T_m). At temperatures considerably below their T_g, polymers are brittle glasses. Very near the T_g, many amorphous materials become somewhat leathery, while above it they go through a rubbery phase before becoming liquid. Elastomers are polymers that are used above their T_g. Like liquids, they change shape or flow easily when a weak force is applied. However, unlike liquids, they return to their original shape after the force is removed. Their polymer chains are very mobile (undergo conformational changes easily) and are held together with periodic crosslinks to prevent permanent deformation.

Semicrystalline polymers with glass transition temperatures below room temperature can be fabricated into tough, flexible plastics. Their chains are flexible because of the low T_g, but the crystallites serve as physical crosslinks that give the material dimensional stability and increase stiffness. Melting semicrystalline polymers produces an amorphous liquid that can be fabricated into useful objects (see Chapter 8). Table 7-2 illustrates this by summarizing the properties of a few specific polymers with a range of glass transition and melting temperatures.

Making chains more linear and more regular (e.g., stereoregular or tactic) so that they can align better enhances crystallinity. Incorporating groups in the polymer that can interact through dipole-dipole, hydrogen bonding, or ionic interactions increases the forces of attraction between chains and can also increase crystallinity. The crystallinity of some polymers can be increased substantially by orienting (stretching) the sample. Synthesizing polymers from all-aromatic monomers that produce chains that can orient with a high degree of order and that have strong interactions leads to high performance polymers such as Kevlar. Polymers such as these can withstand very high temperatures.

Table 7-2. Comparison of properties for polymers with a range of T_g's and T_m's.

Polymer	T_g (°C)	T_m (°C)	Properties
elastomers			
PDMS	-127	-29	liquid until cured, then elastomer
polybutadiene (cis-1,4)	-102	6	liquid until vulcanized, then elastomer
plastics			
HDPE	-125	140	tough plastic
isotactic PP	-35	150	tough plastic
polystyrene	100	—	brittle amorphous plastic
isotactic PS	100	240	tough plastic film when oriented
PVC	85	~285	hard, inflexible plastic; always plasticized

The mechanical properties of polymers can be expressed in a number of ways. Tensile modulus describes the stiffness of a material and provides information about its potential usefulness. Toughness is defined as the work required to elongate a sample to the breaking point. Polymers with high impact strength resist cracking or breaking when struck by a hard object.

Commercial polymeric materials usually contain a number of additives. Plasticizers, fillers, reinforcing agents, impact modifiers, and crosslinking agents are just a few of the many important types.

References Cited

Allcock, H. R., and F. W. Lampe. 1990. *Contemporary polymer chemistry*. 2nd ed., 9. Englewood Cliffs, NJ: Prentice Hall.

Jacoby, M. 2002. Olympic science. *Chemical and Engineering News* February 4: 29–32.

Sperling, L. H. 1986. *Introduction to physical polymer science*. 5. New York, NY: John Wiley and Sons.

Stevens, M. P. 1993. Polymer additives. I. Mechanical property modifiers. *Journal of Chemical Education* 70 (6): 444–8.

Yee, A. F. 1987. Impact resistance. In *Encyclopedia of polymer science and engineering*, ed. J. I. Kroschwitz, 2nd ed., vol. 8, 39. New York, NY: Wiley Interscience.

Other Reading

Campbell, I. M. 2000. *Introduction to synthetic polymers*. 2nd ed. New York, NY: Oxford University Press.

Elias, H-G. 1993. *An introduction to plastics*. New York, NY: Wiley-VCH.

Elias, H-G. 1997. *An introduction to polymer science*. New York, NY: Wiley-VCH.

Grosberg, A. Y., and A. R. Khokhlov, 1997. *Giant molecules: here, there, and everywhere*. New York, NY: Academic Press.

Seymour, R. B., and C. E. Carraher. 1990. *Giant molecules: Essential materials for everyday living and problem solving*. New York, NY: Wiley Interscience.

Sperling, L. H. 1986. *Introduction to physical polymer science*. Chapters 5–9. New York, NY: John Wiley and Sons.

Stevens, M. P. 1993. Polymer additives. II. Chemical and aesthetic property modifiers. *Journal of Chemical Education* 70 (7): 535–38.

Stevens, M. P. 1999. *Polymer chemistry: An introduction*. New York, NY: Oxford University Press.

Walker, F. H. 2001. Fundamentals of polymer chemistry: III. *Journal of Coatings Technology* 73 (914): 67–70.

Section 3 Useful Materials

Chapter 8

Polymer Processing— Making Useful Materials

So far we have examined the essence of polymers, both natural and synthetic. We have looked at a variety of ways that they are synthesized and studied some of their properties. In this chapter we will find out how we can convert a polymer sample that might be in powder or pellet form into some useful object. This is largely the realm of the engineers, who design the processing equipment and determine the conditions that produce polymeric products with optimum properties. These objects take a wide variety of shapes, including films, fibers, solid parts, hollow containers such as bottles, and foamed objects.

The engineers must work closely with the chemists, however, to ensure that the polymer they are processing has the appropriate composition, degree of crystallinity, and molar mass. Materials scientists may also be consulted because the part being manufactured may not consist of pure polymer: It may contain one or more additives or be a composite. The final product must satisfy several important criteria. Its mechanical properties and appearance must remain essentially constant throughout the lifetime of the product. Because most polymers are organic compounds, they are prone to oxidation and degradation. This might only change the appearance of the product (e.g., color fading or the development of a dull surface), or it might lead to a weakening of the product. Therefore the additives often include compounds that help stabilize polymers against degradation by ozone, light, or heat. Pigments might be included in the formulation either to add color or to absorb UV (ultraviolet) radiation or visible light. In this chapter we will examine some of the techniques used to convert polymers into useful objects.

Melt-Processing of Polymers

The simplest technique for converting plastic powder, pellets, or flakes into useful shapes is by melt *extrusion*. An *extruder* is a machine that operates somewhat like an old-fashioned crank meat grinder (see Figure 8-1). Polymer is fed into a chamber where it is melted and mixed with other polymers

Figure 8-1. Main components of a single-screw extruder (courtesy of the Association of Plastics Manufacturers in Europe).

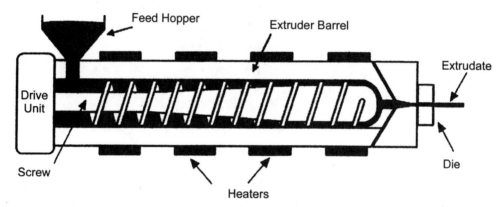

or additives, a process called *compounding*. The molten polymer mixture is then forced under pressure through a *die* as a continuous stream of liquid. In fact, following polymerization, a plastic is often extruded in thin strands that are cooled and chopped into pellets. This is probably the most convenient way for a large chemical manufacturer to handle huge volumes of polymer that need to be packaged and shipped to the many smaller companies that will mix it with other ingredients and fabricate parts.

The tool that performs the mixing and forces polymer through the die is a threaded shaft called a *screw*. Extruders have one or two screws, with the specific design of the screw(s) having an enormous effect on the degree of mixing and the temperatures that are generated inside the chamber. Most of the heat for melting comes from the action of the screw on the polymer, converting mechanical energy into heat energy. The shape of the die determines the nature of the object that emerges from it. A flat die with a number of very small holes (a *spinneret*) produces fiber. A wide die with a narrow slot across its width produces film or plastic sheet. Other dies produce pipe or create insulation on wire. These products are all extruded as a continuous mass having the desired cross-sectional shape and are subsequently cut into convenient lengths. In coextrusion, two or more different polymers are extruded simultaneously to produce a multilayer structure. Approximately 60% of all plastics are processed by extrusion (McCrum, Buckley, and Bucknall 1997).

Injection Molding

Injection molding is a high-volume production technique for turning out thousands of plastic parts per hour. An extruder melts and mixes plastic as above and then forces or injects molten material into a mold that has the shape of the final object. The plastic inside the mold cools, and the mold opens and ejects the part(s). The mold closes, and the cycle repeats. See Figure 8-2. Injection molding enables the mass production of complex shapes

Figure 8-2. Extruder for injection molding (reprinted from Crawford 1987 with permission from Elsevier). a, Mold cavity filled with polymer. b, Mold opened for ejection of solid part.

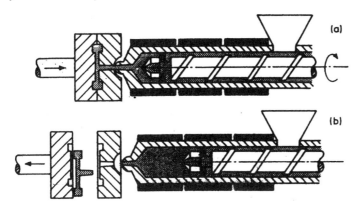

and very small parts, which can be difficult to manufacture by other methods in large volumes. Molds are designed so that the final parts need little or no finishing. Often a number of small parts are formed in the same mold and are joined together by thin plastic strips called *sprues* and *runners*. After leaving the mold, the parts are separated from the sprues and runners, which are waste. Rather than being discarded, this scrap, called *regrind*, is simply reintroduced into the injection-molding machine and remelted. As we will see in Chapter 9, this is called *primary recycling*. Injection molding is actually quite an old technique, the first machine being patented in 1872 for producing parts made from celluloid (Rubin 1986).

Injecting Molding Thermoset Polymers; Reaction Injection Molding

Injection molding can be used with some thermosets in addition to thermoplastics, as long as the process can be controlled such that the crosslinking or curing takes place in the mold and not in the extruder barrel. It has been used effectively, for example, with thermosetting polyester resins. Scrap runners or defective parts must be discarded, however, as they cannot be remelted.

The injection molding process has been significantly simplified in the case of some systems such as polyurethanes. Two or more liquid monomer mixtures are efficiently mixed in the mold, where they undergo step-growth polymerization producing both a polymer and an shaped object simultaneously. In fact, because the polymer is crosslinked, the object consists of essentially one molecule. This process is called *reaction injection molding*, or *RIM*. It has the advantage of collapsing the traditional four polymerization/extrusion/remelting/injection-molding steps into just one step, saving considerable energy. It works best for polymerizations (such as polyurethanes and ROMP [ring-opening olefin metathesis polymerization] of dicyclopentadiene [see Chapter 5]) that are fast, so cycle times in the mold are short. This technique is often used to manufacture large parts for golf carts, personal watercraft, snowmobiles, and automobiles (e.g., bumpers, fenders, spoilers, and fascias).

Figure 8-3. Poly-(ethylene terephthalate) (PET) soda bottle preform or parison.

Bottles and Other Hollow Objects—Blow Molding

Blow molding is an inexpensive and fast method for manufacturing hollow objects like bottles. The technique can be used to produce large parts, too, such as fuel tanks, automobile air spoilers, toy tricycles and wheels, and surfboards. Although a variety of techniques exist for producing bottles, often a *parison* or *preform* is first produced by injection molding. This is a small, thick-walled precursor of the final container (see Figure 8-3). Then, the parison is posi-

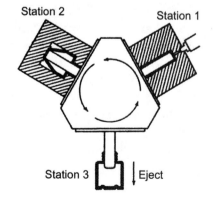

Figure 8-4. Schematic of three-station blow-molding machine (McCrum, Buckley, and Bucknall 1997, courtesy of Oxford University Press). Station 1: injection molding of parison; station 2: blow molding of bottle; station 3: ejection of bottle.

162

tioned in a mold the size and shape of the final bottle, heated to soften the plastic, and injected with air so that it conforms to the shape of the mold. On cooling, the bottle is ejected and the cycle repeats. The injection and blow-molding operations are often consolidated on one machine (see Figure 8-4).

Bottles range in quality from cheap to pretty sophisticated, depending upon the use to which they are put. The polymer(s) used in the construction of the bottle is chosen based on strength, compatibility with the contents, clarity, and cost. We would prefer that the bottle not break if dropped on the floor. The marketing people tell us we prefer certain products, especially foods, in optically clear bottles. Plastic bottles are, after all, primarily a replacement for glass. Often a bottle not only must contain its contents but must also prevent the passage of gases from one side of the bottle to the other. In other words, the *permeability* of the bottle and of the polymer(s) from which it is made need to be taken into consideration. Finally, only certain polymers are approved for contact with food.

For example, consider a soft drink or soda bottle. What are the requirements? Cheap! Strong enough to contain up to 2L of soda. Optically clear. And very important, a barrier for CO_2, because carbonated beverages taste "flat" once the concentration of carbon dioxide falls too low. Table 8-1 lists the permeability of various polymers to CO_2 as well as to oxygen and water vapor.

Table 8-1. Relative permeabilities of sample polymers for three key gases (Salame 1986; Pauly 1999).

Polymer*	CO_2	O_2	H_2O
LDPE	1025	116,000	5.6
polystyrene	650	104,000	61
polycarbonate	650	56,000	89
isotactic polypropylene	750	36,000	3.3
HDPE	30	26,400	1
poly(methyl methacrylate)	—	4000	83
PVC (unplasticized)	15	2600	14
PET (bottle grade)	15	1200	16
poly(vinylidene chloride)	2.5	60	1.1
ethylene-vinyl alcohol copolymer	4.0	78	
poly(vinyl alcohol)	1	1	2200

*The higher the number, the greater the degree of permeability.

As we can see from the table, PET (polyethylene terephthalate) is a relatively good barrier to CO_2, making it a likely candidate for constructing a soda bottle. Furthermore, it has been approved in many countries for contact with food, it is relatively inexpensive, and it has sufficient strength (we have already seen that it forms strong fibers). In addition, PET is a semicrystalline polymer. And it is this property that makes PET particularly useful for soda bottles. Although we can blow mold bottles of PET that would be strong enough for soda bottles, they would be relatively thick, too flexible, and relatively heavy. However, if we *stretch* the bottles while blow molding them, just above T_g but below T_m, we will *orient* the polymer in two dimensions and generate a number of microscopic crystallites. As we have seen previously (Chapter 7), orientation strengthens the polymer and makes it stronger and stiffer while making it thinner. After *stretch blow molding*, the bottle is quickly cooled below T_g to lock in the structure. The result is a thinner, lighter bottle that is stronger and a little more rigid. It will be easier to pick up and pour from. In addition, it will prevent the loss of CO_2 for several weeks or months (depending upon the size of the bottle), assuring that the soft drink will taste just like it is supposed to. The polymer chains in such bottles are under considerable strain. The strain can be relieved by reheating the empty bottle to a temperature near or above T_g, allowing the chains to relax. At this point, the bottle will shrink to approximately one-third its size. See Section 4 for a specific exercise.

Multilayer structures can be blow molded also, an operation analogous to coextrusion. Sometimes one particular polymer is not totally suitable for a specific application. Polyethylene and polypropylene are approved for contact with food and are inexpensive. As a result, both polymers are used for packaging many food items. Although both polymers are effective barriers to water vapor, as we can see in Table 8-1, both are quite permeable to oxygen, which would shorten the shelf life for oxygen-sensitive products. On the other hand, copolymers of ethylene and vinyl alcohol (EVOH) are excellent oxygen barriers. The relative diffusion rates of O_2 through equal thicknesses of LDPE (low-density polyethylene) and EVOH is 29,000/1. Therefore making bottles with a sandwich structure results in a product with all of the desired features. The most likely structure would place polyethylene (or polypropylene) on the inside and outside, with a thin layer of EVOH copolymer in the middle. If two different types of polymers in a multilayer structure are very dissimilar, they might tend not to stick to each other very well. To solve this, a thin adhesion layer is placed between them. Thus a three-layer bottle would actually require five layers to produce a practical container (Figure 8-5).

As we will see in Chapter 9, multilayer blow molding is sometimes used as a way to incorporate recycled polymer in food-grade bottles. For safety reasons the use of recycled plastics in packaging that comes in direct contact

Figure 8-5. Schematic cross section of a "three-layer" bottle.

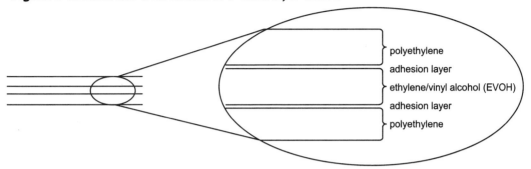

polyethylene
adhesion layer
ethylene/vinyl alcohol (EVOH)
adhesion layer
polyethylene

with food is banned. Recycled containers can be cleaned, shredded, and re-processed, and then sandwiched in the middle of virgin polymer so that the food product does not contact the recycled polymer.

Cast and Blown Films

In Chapter 6, we discussed one method of preparing polymer films using *solution casting*. Now we will consider melt casting, a process we mentioned earlier in this chapter. In this process, polymers are melted and forced through thin, wide dies to produce films or sheets. Film is generally produced in one of two ways. In *cast film*, molten polymer emerges from a die onto a temperature-controlled roller, forming a continuous, single layer of film that cools at a rate adjusted to control crystallinity. This process is shown schematically in Figure 8-6. Following removal from the casting roller, the film is often oriented by stretching it. Films produced by this process range from approximately 0.5 to 20 mil (~0.15 mm – 0.5 mm) (1 mil is 0.001 inch or 25 mm). Examples include PET X-ray film support and LDPE paint drop cloths.

Figure 8-6. Schematic of cast film extrusion process (reprinted with permission from Tess and Poehlein, copyright 1985, American Chemical Society).

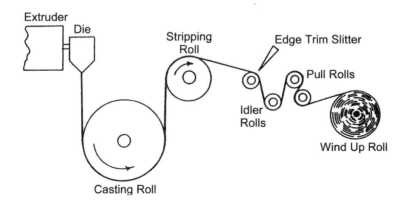

Products much thicker than 0.5 mm are usually considered to fall in the category of sheet rather than film.

The other major process produces *blown film*, a continuous sleeve of thin polymer. The schematic for this process is in Figure 8-7. Rather than having a long, horizontal slit that produces a flat width of film, the die for blown film has a circular opening that produces a column of film. The column or sleeve is sealed at the top and expanded with air. At the top, the column is collapsed into a double layer of film. This is either rolled up as is, or is slit into single sheets. Just as in cast film, orientation is possible here also. Most plastic bags and many flat film products found around the home are made from blown film (e.g., trash, lawn debris, grocery, produce, and dry-cleaner bags; shrink film; and poly[vinylidene chloride] kitchen self-cling wrap).

Figure 8-7. Schematic diagram for film blowing process.

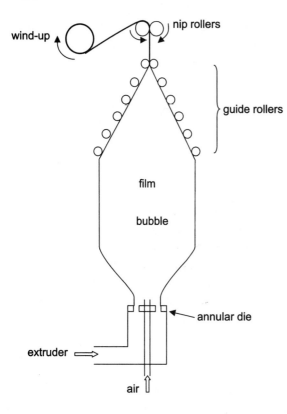

Both the cast and blown films can be coextruded to produce films consisting of more than one layer, allowing special effects and the combining of properties of different polymers. For example, if a high-strength film is needed for a particular application, the manufacturer might be able to save considerable money using coextrusion. For this, a thin layer of a high-performance (expensive) polymer might be sandwiched between two inexpensive commodity polymers rather than extruding a thicker, single layer of the high-performance polymer.

Gore-Tex

What is Gore-Tex and how does it work? Skiers, hikers, kayakers, and others who enjoy outdoor activities are familiar with its ability to keep liquid water out while allowing water vapor to pass through. Gore-Tex is actually a membrane composed of a thin sheet of poly(tetrafluoroethylene) (PTFE) that has been stretched to create submicrosized pores thousands of times smaller than raindrops yet hundreds of times larger than a water molecule. As seen in the photograph, PTFE fibrils are separated from one another on stretching, creating billions of tiny openings. The membrane is almost always laminated to an outer polyurethane layer and an inner liner to create a breathable fabric that sheds water. If used as an article of clothing, water vapor (perspiration) tends to pass from the warmer inside to the cooler outside. This means that waterproof boots and rain gear will provide protection against the elements while keeping the wearer from becoming sweaty.

Membranes such as Gore-Tex make useful filtration media, also. The tiny openings can be engineered to trap particles of various sizes, air pollutants, chemicals, and bacteria.

Gore-Tex is but one of a number of products produced by W. L. Gore and Associates, Inc., a company founded in 1958 by Wilbert and Genevieve Gore to find uses for fluorinated polymers. Gore-Tex was discovered by Wilbert's son, Robert in 1969.

Thermoforming

In *thermoforming*, one of the oldest processing methods, a sheet of thermoplastic is heated above its softening temperature and then forced into a mold. This is accomplished by placing a matching mold on top of the sheet, by using air or vacuum, or by some combination of these. The technique is used to make trays, drinking cups, disposable dishes, food storage containers, storage bins, luggage, signs, and a number of other objects. Sometimes the thermoforming operation is integrated with sheet extrusion, enabling energy savings and boosting efficiency. This is important for extremely cheap parts, such as coffee cup lids, which one machine can produce at a rate in excess of 100,000 per hour. The process is shown schematically in Figure 8-8.

Figure 8-8. Schematic of thermoforming process using matching molds.

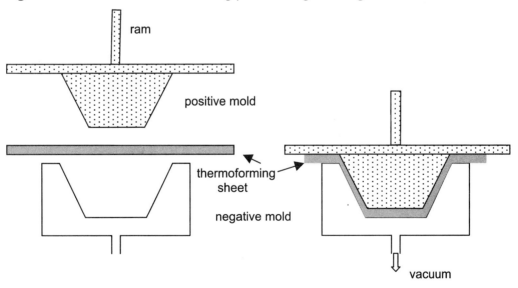

Foamed Objects

A *foam* is a blend of a polymer with a gas, usually air. The foam is normally formed during processing, when a *blowing agent* releases a gas that becomes trapped inside the polymer matrix as small bubbles. For a thermoplastic, cooling below T_g stabilizes the foam. For an elastomer, crosslinking following the generation of the foam provides the necessary stabilization. Examples are foamed rubber and polyurethane foams. The latter are designed to be either hard and rigid, useful for thermal insulation applications, or soft. Soft polyurethane foams are used for cushioning in upholstered furniture and in automobiles. A foam can be extruded, producing a range of items such as foamed polystyrene insulation board, pipe, and packaging material.

One way of forming a foam is through the use of a volatile liquid as the blowing agent. A low-boiling hydrocarbon such as pentane (b.p. 36°C) can be forced into small polystyrene beads, say, with heat and pressure, and then cooled. On being reheated at atmospheric pressure, the pentane turns to a gas, expanding the PS beads to several times their original size. When this step is carried out in an extruder, the temperature is high enough so that the expanded beads stick together when the foam emerges from the die. The pentane vaporizes, leaving air on the inside of the beads. The product in this case is *expanded polystyrene foam*, or *EPS*. Thin sheets of EPS are commonly converted into insulated drinking cups by thermoforming. Close examination of EPS products will show a series of fused beads rather than a continuous matrix of polymer containing little cells of gas. Coextrusion can be used to produce cups with different colors on the outside from that on the inside. Other common products made from EPS are clamshell food containers, egg cartons, meat trays, foam coolers, and insulation board.

Fibers

Fibers are used in a wide variety of applications and are composed of diverse materials. Natural animal and plant fibers, including wool, silk, cotton, hemp, ramie, flax, jute, and sisal, have long been used for clothing, baskets, fishing nets, and rope. Ropemaking has been an essential skill for thousands of years, a key element in the advancement of civilizations via the oceans. The natural mineral fiber asbestos is in the silicate family. We have encountered glass fibers previously in the text (see Chapter 7).

The first man-made organic fiber was produced in the midnineteenth century by converting cellulose into soluble cellulose nitrate and then forcing the solution through small holes (spinnerets) to form fiber. Although this highly flammable fiber was not very practical, it began a series of discoveries that led to Chardonnet's 1885 patent for a rayon fiber produced by regenerating cellulose from cellulose nitrate fiber (viscose process; see Chapter 4). An entirely new industry was launched that today manufactures more than 30 million tonnes of man-made fiber a year, which translates to a worldwide

Figure 8-9. Worldwide textile fiber production from 1950 to 2000 (Rebenfeld 1986; Anon. 2001).

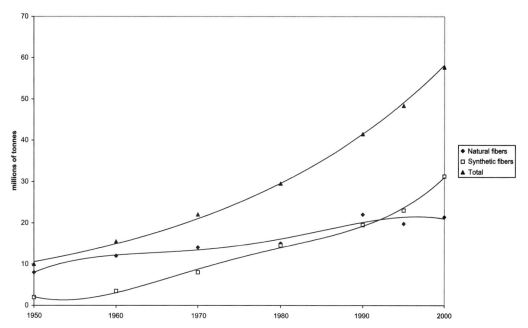

consumption of more than 5 kg/person, or more than 15 kg/person in North America (more than 33 lb/person) (Anon. 2001). Figure 8-9 shows worldwide fiber production for natural and synthetic textile fibers from 1950 to 2000. Total production has increased in response to the world's population growth, from 2.5 billion people in 1950 to 6.8 billion in 2000. Note that the production of synthetic fibers has increased substantially faster than that of natural fibers, surpassing the latter in the early 1990s. In fact, it is apparent that natural fiber production actually reached a plateau in the last decade. As we saw in Chapter 4, the production of synthetic fiber requires very little land. Therefore, substituting synthetic fiber for natural fiber frees enormous areas of arable land for food production.

What do we think of when we hear "fiber"? Clothing, certainly, and other textiles such as sheets and blankets, curtains and upholstery. Some fiber goes into carpeting—for offices, homes, automobiles, and recreational vehicles. Automobile, truck, and bus tires are strengthened with the use of fibers called *cords*. Some fiber is used for industrial purposes such as insulation and filtration. Hollow fibers that act as membranes are used in the desalination of seawater by reverse osmosis and in kidney dialysis. As we have already seen, extremely high-strength fibers are used to make bulletproof safety equipment and to reinforce polymers for high performance applications such as aerospace and sports equipment. In medicine, fibers have long been used for sutures, dressings, and tapes. *Engineering fibers* can be fabricated into porous or hollow structures that serve as synthetic blood vessels, artificial ligaments, or scaffolds for tissue regeneration. *Optical fiber* has become a crucial technology for the rapid transmission of data and for telecommunications. The single largest use for fibers, however, is for the manufacture of paper and cardboard. Most of this is natural cellulose, but synthetic papers are increasingly popular (see sidebar that follows).

Tyvek

What is that white material that often covers new or remodeled buildings under the outer layer of brick, vinyl, or wood siding? The name Tyvek is often prominently printed on it, as is the name of its manufacturer, DuPont. Tyvek is the name given to a family of products, all constructed of high-density polyethylene (HDPE) fibers that are interconnected and then bonded together with heat and pressure (see photomicrograph). Grades of Tyvek in which fibers seven times thinner than a human hair are densely packed look and feel like ordinary paper. These are used in envelopes, packaging, and waterproof maps and trail guides. Unlike paper, which is made from cellulose fibers, plastic "paper" is very tough and is difficult to tear. Other grades resemble fabrics and are sewn into protective clothing for clean-room operations or chemical and drug manufacturing.

Heavier grades are used in construction and for covering boats, RVs, and snowmobiles. These products provide abrasion resistance and protection from water, dirt, dust, and sunlight. In addition, they "breathe," meaning that water vapor and air can pass through rather than being trapped on the inside.

Photomicrograph of Tyvek courtesy of Dupont.

With the exception of silk, which the silkworm or spider extrudes as a continuous filament, natural fibers are of finite length. For textile use, these need to be cleaned and then *spun* into threads or yarns. Synthetic fibers, on the other hand, are continuous filaments produced from a solution or melt. The term "spinning" is used to describe the formation of synthetic fibers, but in this sense it has no relation to the process for combining fibers into threads.

For *melt spinning*, a thermoplastic is first heated above its melting point (often in an extruder). Then the molten polymer is filtered and forced through a spinneret. Fibers emerging from the spinneret are cooled under controlled conditions, passing over guides and rollers to a take-up spool or bobbin. Often a *finish* is applied before windup to control static electricity and friction (Stevens 1993). Large-scale production machinery produces fiber at a rate of thousands of feet a minute. A schematic diagram of a melt-spinning apparatus is drawn in Figure 8-10.

Some polymers, such as aramids and cellulosics, have such high melting temperatures that they thermally decompose before they melt. Fibers from these high-melting materials are produced by *solution spinning*. Rather than molten polymer, it is a polymer solution that is forced through the spinneret. The filaments emerge from the die into a different solution called a coagulating bath, where the polymer is either precipitated or regenerated (e.g., rayon).

Filaments of most fibers are *drawn*, meaning they are stretched following spinning the same way that films are stretched. This step increases crystallinity and orients the molecules with respect to the fiber axis. The result is a much stronger fiber. Nylon and polyester fibers, for example, which have little molecular orientation after spinning, are stretched 4 to 6 times their original length (a draw ratio of 4 to 6). For a typical nylon, the tensile modulus triples at a draw ratio of 6. In modern production, spinning and drawing are combined in one operation.

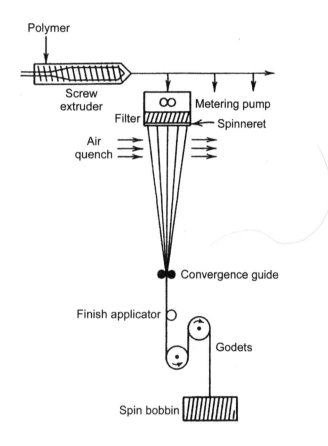

Figure 8-10. Schematic diagram for the extrusion and melt spinning of polymeric fibers (McIntyre and Denton 1986, reprinted by permission of John Wiley and Sons, Inc.).

What properties must a polymer have to serve as a strong fiber, an unusual object because most of its strength is in only one dimension? For one, the material has to have high tensile strength and a high elastic modulus (high stiffness). Generally the polymer chains are linear, have a high degree of symmetry, and are highly crystalline. Some degree of flexibility (e.g., in the amorphous regions) is also helpful, as this increases the ability of the fiber to stretch under tension. Most of the strongest fibers have strong attractive forces between molecules, such as hydrogen bonding. Many fibers are made from step-growth polymers such as polyesters and polyamides (nylons) because these types of polymers often possess most of the required characteristics.

Fibers used in textiles have special needs. For those that need ironing, the T_m must be at least 200°C. If the crystallites in the polymer were to melt, the fibers would lose their reinforcement and would shrivel. It is desirable that the T_g be significantly above room temperature so that intentional creases put into a fabric by ironing remain there during wear. Those fabrics that will be dry-cleaned must have a high degree of solvent resistance. The ultimate properties of fibers depend a great deal upon their method of manufacture. Different treatments during and after spinning determine, for example, how the fiber reflects light, how it feels, how flammable it is, how easily it is dyed, and how much static electricity it builds up.

Some fibers are used as monofilaments (single fibers), for example, as fishing line, as bristles for brushes, or as classical guitar strings. Most of the fiber produced, however, is combined with other fibers into threads or yarns. Not all fibers are highly crystalline and stiff. Some are designed to be elastomeric, such as rubber or spandex. The latter (e.g., DuPont's Lycra) is a polyurethane copolymer that is prepared with alternating segments of *soft* (low T_g) elastomeric repeat units and *hard* (high T_g or crystalline) repeat units. The soft segments typically constitute approximately 80% of the total mass of the polymer. Just as we saw when we discussed thermoplastic elastomers, the hard segments phase-separate from the soft ones during processing, resulting in rigid sites that act as crosslinking points, preventing permanent deformation of the soft segments during elongation. See "Spandex—Not So Firm a Foundation" in Chapter 7.

Specific applications require fibers with the right combination of properties. For fibers that are used in engineering applications, strength and mass are the most important considerations. Steel is commonly fabricated into cable for suspension bridges, for guylines for towers, and for nautical anchor or mooring lines, among other applications. It corrodes (rusts), is an electrical conductor, and is quite heavy. Ropes and cables can be constructed of engineering fibers that have superior strength but at a fraction of the mass of steel. Aramid is one example, as is ultrahigh molar mass polyethylene. This is a relatively new fiber that is spun from solution (as are aramids), allowing the extremely long chains to orient very efficiently and form fibers that are stronger than aramids and several times stronger than steel.

Table 8-2 summarizes the relative strength, relative elastic modulus, and extension at break (percent elongation when the fiber breaks) for a variety of common fibers. Also included are a few properties and some typical applications.

Table 8-2. Summary of properties of different types of monofilament fibers (Rebenfeld 1986; Smit, Jacobs, and van Dingenen 2003).

Fiber	Relative Strength	Relative Elastic Modulus	Elongation (%)	Properties	Typical Applications
common fibers					
polypropylene	1.8	1.1	30	light, hydrophobic	wicking sports apparel, rope
nylon	1.6	1.0	30	absorbs loads elastically	carpets, tire cord, apparel, fishing line
polyester	1.2	2.2	30	high elastic recovery	apparel (e.g., "fleece"), carpets
acrylic (acrylonitrile)	1.0	1.8	20	stress-strain behavior similar to wool	blended with wool for apparel, carpets
engineering fiber					
aramid (Kevlar)	4.5	20	4	high strength, stiffness	tire cord, advanced composites, ropes and cables, body armor
ultrahigh molar mass polyethylene	5.7	28	3	very high strength, excellent water, salt water resistance	cut-resistant gloves, vehicle and body armor, ropes and cables, fishing line
elastic fiber					
spandex	0.2	0.001	600	recovers from large extensions	sports apparel

Summing Up

We have covered considerable ground in this chapter, discussing methods used to produce a wide array of polymer products. For any of these products to be useful, the polymers used in fabricating them must possess a set of properties that is consistent with the use to which the product will be put. Perhaps it would be helpful to summarize some of the basic properties typically found in objects made with different types of polymers. Table 8-3 describes some typical properties of different classes of materials, including the presence or absence of crystallinity, chain architecture, and use temperature.

Table 8-3. Comparison of key properties for practical applications (Sperling 1986).

Polymer Type	Morphology	Architecture	Use Temperature	Example
elastomers	amorphous	crosslinked	above T_g	rubber band
adhesives	amorphous	linear or branched	above T_g wet; below T_g dry (sticks)	poly(vinyl alcohol) lickable postage stamp, label
plastics	amorphous	linear or branched	below T_g	PMMA, PS
	semicrystalline	linear	slightly below to above T_g; below T_m	HDPE, PP
fibers	semicrystalline	linear	sl. below to above T_g; below T_m	nylon, polyester, aramid
coatings	amorphous	crosslinked	near T_g: flexible but not rubbery	house paint

For example, poly(vinyl alcohol) (PVOH) finds application as an effective adhesive for paper products such as the old-fashioned postage stamps or envelopes. Licking the stamp or envelope flap causes the adhesive to become sticky because water plasticizes the PVOH and lowers the T_g below room temperature. On pressing the stamp to a piece of paper, the polymer chains can entangle and hydrogen bond with the cellulose in the paper. As the water evaporates, the T_g rises again above room temperature, polymer chain motion effectively ceases, and the bond is set. At least until the envelope gets wet! This is only one of a number of types of adhesives, different types working by different mechanisms.

Thermoplastics are often compounded in an extruder, a machine that heats the polymer, mixes it with other ingredients, and then extrudes it through a die. Many parts are made by injection molding or injection blow molding. Films can be cast as a single continuous sheet or blown into a continuous column of polymer. Stretching and orientation are often important parts of the fabrication process and can affect properties of the final product, such as clarity, solvent resistance, and permeability to various gases. Multiple layers can be produced by the process of coextrusion. Fibers are extruded through spinnerets and are often drawn and oriented to increase strength.

References Cited

Anon. 2001. Manmade fiber production tops 31 million tons as natural fibers post slight decline. *International Fiber Journal* 16 (4): 16–26.

Crawford, R. J. 1987. *Plastics engineering*. 2nd ed., 187. New York, NY: Pergamon Press.

McCrum, N. G., C. P. Buckley, and C. B. Bucknall. 1997. *Principles of polymer engineering*. 2nd ed. New York, NY: Oxford University Press.

McIntyre, J. E., and M. J. Denton. 1986. Fibers manufacture. In *Encyclopedia of polymer science and engineering*, ed. J. I. Kroschwitz, 2nd ed., vol. 6, 809. New York, NY: Wiley Interscience.

Pauly, S. 1999. Permeability and diffusion data. In *Polymer handbook*, eds. J. Brandrup, E. H. Immergut, and E. A. Grulke, 4th ed., VI/543–VI/569. New York, NY: John Wiley and Sons.

Rebenfeld, L. 1986. Fibers. In *Encyclopedia of polymer science and engineering*, ed. J. I. Kroschwitz, 2nd ed., vol. 6, 647–733. New York, NY: Wiley Interscience.

Rubin, I. I. 1986. Injection molding. In *Encyclopedia of polymer science and engineering*, ed. J. I. Kroschwitz, 2nd ed., vol. 8, 102. New York, NY: Wiley Interscience.

Salame, M. 1986. The use of barrier polymers in food and beverage packaging. *Journal of plastic film and sheeting* 2(4): 321–34.

Sperling, L. H. 1986. *Introduction to physical polymer science*. 290–91. New York, NY: Wiley Interscience.

Stevens, M. P. 1993. Polymer additives. III. Surface property and processing modifiers. *Journal of chemical education* 70(9): 713–18.

Tess, R. W., and G. W. Poehlein, eds. 1985. *Applied polymer science*. ACS Symposium Series, 285 2nd ed., Washington, DC: American Chemical Society.

Other Reading

Elias, H-G. 1993. *An introduction to plastics*. Ch. 10 and 11. New York, NY: Wiley-VCH.

Elias, H-G. 1997. *An introduction to polymer science*. Ch. 11–14. New York, NY: Wiley-VCH.

Stevens, M. P. 1999. *Polymer chemistry: An introduction.* 3rd ed., Ch. 1, 4. New York, NY: Oxford University Press.

Tess, R. W., and G. W. Poehlein, eds. 1985. *Applied polymer science.* ACS Symposium Series 285, 2nd ed., Washington, DC: American Chemical Society.

Disposal, Degradation, and Recycling; Bioplastics

Everyone is familiar with plastic waste. We throw away large volumes of it, at home, at school, at work, at fast food restaurants, on vacation. Much of it ends up in the trash. We see some of it as litter along the sides of roads, streams and lakes, and floating up on beaches. Why does it stick around so long? Some polymers do degrade when left in the sun. Why don't more of them? We probably recycle some used plastics, although how much depends upon where we live. In many localities, only items produced from PET (polyethylene terephthalate) and HDPE (high-density polyethylene) are collected for recycling. Why don't we recycle more of it? Why not LDPE (low-density polyethylene) and polystyrene? And what happens to it when we do? We'll develop some basic principles in this chapter on some of the avenues that help us follow the U.S. Environmental Protection Agency's advice to "reduce, reuse, recycle."

The "Polymer Cycle"

In biology you've probably studied the carbon, oxygen, and nitrogen cycles. The carbon and oxygen cycles describe the interplay of different cells and organisms in nature in the utilization of CO_2 and water from the atmosphere for the production of glucose and oxygen by photosynthesis. The other half of the cycle describes the metabolic oxidation of glucose to CO_2 and water. Could we consider constructing a *polymer cycle*? To do that, we would need to know where polymers come from and what happens to them when they are discarded. This might be a good time to review where polymers come from, anyway.

First, it's pretty clear where natural polymers originate. When organisms die, the polymers in them eventually break down and are returned to the earth. The cellulose in dead trees and plants is hydrolyzed by microorganisms, the products metabolized and returned to the atmosphere as CO_2 and water. Photosynthesis in green plants converts this CO_2 into new cellulose in living trees and plants, and the cycle continues.

Proteins are likewise broken down by microorganisms, or eaten by mammals and hydrolyzed by enzymes to amino acids. These are further oxidized to NH_3, H_2O, and CO_2, releasing energy. These inorganic compounds can be utilized by living organisms to synthesize proteins, or they can be excreted. Bacteria convert ammonia to nitrites and nitrates, which plants can convert to amino acids.

How about synthetic polymers? Remember that most, but not all, come from petroleum (recall Figure 1-1). Some 90% to 95% of crude oil is refined into heating oil, or into gasoline, diesel, and jet fuel, and burned in internal combustion engines. Less than 10% of crude oil is refined to produce organic *petrochemicals*, approximately half of which end up as monomers. Other uses of petrochemicals include solvents and small-molecule organic compounds that are the building blocks for drugs, detergents, antioxidants, fertilizers, dyes, vitamins, pesticides, and many other applications. The mono-mers are polymerized to make polymers, which are then processed into prod-ucts, such as fibers, elastomers, and coatings. And of course, plastics. We'll call the conversion of petroleum into polymers and their subsequent pro-cessing into useful objects the *production* phase of the polymer cycle. The next phases are *use* (or *life*), followed by *disposal* (or *end-of-life*). This is summarized in Figure 9-1.

Figure 9-1. The production, use, and disposal of polymers.

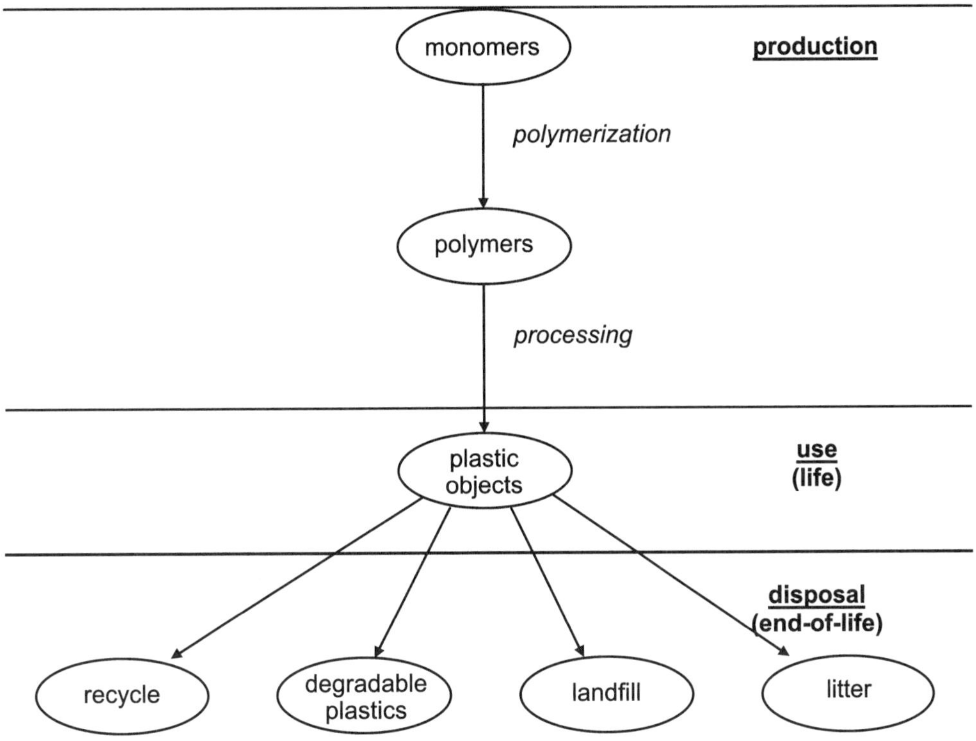

The Disposal Phase: What Are Our Options?
Just Throw It Away!

We buy these products, *use* them (sometimes for many years, sometimes for only a few seconds), and then look for a place to *dispose* of them. For some, the nearest roadside, public park, beach, or body of water works quite well. Fortunately, a combination of statutes, public initiatives, and education over the last 40 years or so has significantly reduced *litter*.

Until fairly recently, many merchant ships around the world would dump their refuse into the ocean rather than bring it back to port. Much of it floated, posing hazards to smaller boats, and would eventually wash up on beaches. Not only was this undesirable from an aesthetic standpoint, some of the litter—such as used medical syringes—posed real hazards. International laws now make this practice illegal, somewhat reducing the abuse. Sometimes, however, litter is the result of an accident rather than a deliberate act. Storms at sea routinely cause containers of products to wash overboard, break open, and distribute their contents into the ocean. In addition, each year a significant amount of fishing line and netting is lost at sea, with the result that many thousands of seals and many hundreds of thousands of seabirds become trapped in it and die.

Some unfortunate events, however, have positive outcomes. In 1990, the cargo ship *Hansa Carrier* from Korea lost 21 boxcar-sized shipping containers during a storm. Of the containers that broke open, four of them were filled with more than 61,000 Nike sneakers, many of which washed up on beaches in the state of Oregon. The number was so large that beachcombers held swap meets so as to match lefts with rights of the same size (Krajick 2001). In 1992, 29,000 bathtub toys were accidentally released in the northern Pacific Ocean and eventually washed up on the beaches along some 500 miles of Alaska's coast. Interestingly, the sneakers and the rubber duckie litter provided a unique scientific opportunity. Oceanographers Curtis Ebbesmeyer and James Ingraham used information from both of these accidents to better understand long-range ocean currents and to refine a computer model predicting the currents based on weather events (Krajick 2001).

Because of their low densities, most of the plastic objects released at sea remain on the surface, where they slowly break apart into smaller and smaller pieces. In fact, the surface of the ocean contains a large volume of plastic refuse, the most common man-made objects sighted. Much of it is in the BB-shot-to-fingernail size range. This phenomenon is significant enough that the term *nurdle* was created to describe these small pieces of plastic flotsam. Some of the nurdles are mistaken for food by animals such as turtles, fish, whales, and birds, whose stomachs often contain sizable amounts of plastic trash (Krajick 2001). One study estimates that the near-shore water off the beaches near Los Angeles contains more than one million nurdles per mile.

Better Yet, Bury It

Back on land, most of our discarded products go into the trash, which for many of us is picked up once a week and hauled to a *landfill*. There the kitchen garbage, newspapers, cardboard boxes, clothing, furniture, metal objects, and polymers are buried in an atmosphere depleted in oxygen (*anaerobic* conditions). Although this concept is very much misunderstood, it is important to realize that, in a landfill, *degradation* occurs extremely slowly. Newspapers, plastics, metals, food (e.g., ears of corn) change very little over a period of many years (Rathje 1989). Even polymers designed to degrade will not do so effectively when buried in a landfill (Scott 1995). The lack of degradation in a landfill is actually good, because to do so in anaerobic conditions would release methane gas, posing an explosion hazard. Furthermore, extensive degradation would cause the landfill to sink (subside). Liquid produced during the degradation, called *leachate*, would seep down into the groundwater. In older landfills, the leachate would include a number of toxic chemicals from newsprint, motor oil, paint, pesticides, and household cleaning materials. Although no one wants one in his or her own backyard, a properly constructed and maintained "sanitary" landfill is not an unreasonable place in which to bury some of our municipal waste. Government regulations, oversight, and newer construction techniques reduce many of the environmental problems that are associated with older, poorly designed landfills, such as hazardous liquids leaching into the groundwater, odors, rodents, and smoldering fires. In the United States, approximately 62% by mass of municipal solid waste ends up in landfills (Rader 1995). Of this, approximately 9% by mass is plastics. An unknown, but small, amount of waste ends up as litter. So why do the nurdles at sea and the plastics in the landfill require many, many years to break down? Perhaps this would be a good time to better understand the concept of degradation.

How Degrading!

Most of us (not all, obviously!) dislike litter. But what is litter? Actually it is really hard *not* to see. As we walk through the forest, litter surrounds us. Trees and plants die and accumulate on the ground. Periodically we encounter the remains of some deceased animal, perhaps sensing its presence with our nose long before we see it. Each fall, deciduous trees shed their leaves, littering our lawns with large volumes of dead vegetation. Shells, seaweed, and the bodies of dead critters wash up on the beach. This is all litter, but we often don't think of it as such. It is part of the natural scene. Yes, dead fish accumulating at our favorite beach spot can spoil an otherwise fine day. But all of these are natural events, at least for the most part. There is beauty in a huge fallen tree deep in the forest covered with many varieties of green mosses and lichens. Who can resist picking up a beautiful seashell? We object more to man-made litter, because it does not blend into the natural environment and because we have the power to do something about it.

An important point, obviously, is that the natural litter will disappear over time. The leaves, if left in place, break down and return organic matter to the soil. In the meantime, they may be a nuisance. Many individuals and some communities collect organic litter and *compost* it, producing an organic-rich medium useful in gardening. That tree in the forest will slowly disappear, as small critters and microorganisms use it for housing and for food. It is part of the food chain, a seedbed for new growth. This is all part of the natural cycles: carbon, oxygen, and nitrogen.

Glass bottles, aluminum cans, or plastic containers left by the side of the road will not decompose or degrade in any reasonable lifetime. Unless picked up, they will stay there for many, many years, reminding us, perhaps, of someone's excess or carelessness. In our view, they don't belong there.

Why don't most synthetic polymers readily degrade? And what is degradation, anyway? Can we make polymers that do degrade? We'll try to answer these and other questions in this section. Scientists often define *degradation* as a negative "change in the chemical structure, physical properties, or appearance of a polymer" (Wool 1995). We have already encountered this when discussing polymer processing. Each time a plastic is melted and extruded, the molar mass decreases, lowering physical properties. This is the result of *thermal* and *mechanical degradation*. This is why engineers sometimes add antioxidants and stabilizers during processing.

But what happens in the environment? A piece of high-density polyethylene from, say, a half-gallon milk container, if left on the ground, will eventually become brittle, and its top surface will become rough and eroded. This is the result of oxidation brought about by exposure to air and sunlight, not because of attack by microorganisms. Microorganisms cannot utilize most synthetic polymers until the polymers have been degraded to small-molecule metabolites that can be absorbed by the organism's cells. Therefore this preliminary period of environmental oxidation is called *abiotic degradation*. For many plastics, especially those to which antioxidants and stabilizers have been added, this can take years.

Step-growth polymers, such as polyesters, can degrade by hydrolysis, a reaction that we have discussed previously. So why do PET soda bottles last so long when discarded by the side of the road? As we have seen, much of the strength of the bottle is the result of microcrystallinity. This same crystallinity that leads to desirable physical properties also helps stabilize PET against environmental hydrolysis, prolonging its degradation. Perhaps this is good, or else bottles filled with cola beverages (pH ~2) could begin to hydrolyze and degrade before their contents were consumed!

Many polymers are specifically designed *not* to degrade in the environment—outdoor signs, deck furniture, boats, automobile bumpers, tires, windshield wipers, and outdoor carpeting to name a few.

A Chain Is No Stronger Than Its ...

So what can we do to speed up the disintegration of polymers that we want to degrade when exposed to the atmosphere? One approach is to use a *natural filler* that is attacked by microorganisms. For example, starch beads can be blended into polyethylene film, significantly increasing the rate of environmental degradation of the film. Microbes remove the starch, leaving holes, which weaken the film. The process can be accelerated even more by adding species (e.g., transition metal compounds) that catalyze oxidation of the plastic. These products have been used in agriculture as plastic mulch. The films are laid down in the fields in rows to prevent weeds from growing, to retain moisture, and to raise the soil temperature for certain crops. The starch-filled polyethylene film degrades either before harvest or before the field is plowed again for the next crop.

Another approach to degradable polymers is by *deliberate synthesis*. By copolymerizing a monomer that causes instability, it is possible to enhance degradation. This is somewhat analogous to inserting periodic weak links into a chain. Of course, it is important to control the process so that the polymer object serves its useful function before degradation begins. Consider the ring or loop carriers that package beverage-can "six-packs." They need to be strong and somewhat elastic. No one wants a pressurized soda or beer can to fall out and break open on the floor or in the car. Polyethylene is a good choice for an application such as this, given its toughness. As litter, these 6-pack rings pose special difficulties. Both land and sea animals can become caught in the rings and die.

Although rings made of polyethylene work very well, as we have already found, polyethylene degrades extremely slowly in the environment. A small amount of carbon monoxide (CO) can be polymerized with ethylene to produce a copolymer that degrades in sunlight (undergoes *photodegradation*):

Equation 1

$$CH_2{=}CH_2 \quad + \quad CO \quad \longrightarrow \quad \left(\!\!CH_2{-}CH_2\!\!\right)_m\!\!\left(\!\!\overset{\displaystyle O}{\overset{\displaystyle \|}{C}}\!\!\right)_n$$

In practice, the CO units constitute less than 5% of the total polymer. Figure 9-2 shows the very rapid decrease in molar mass for an ethylene–carbon monoxide copolymer exposed to a constant source of UV radiation (Harlan and Kmiec

Figure 9-2. Decrease in molar mass for low-density polyethylene and ethylene-carbon monoxide copolymer on exposure to UV radiation (Harlan 1995).

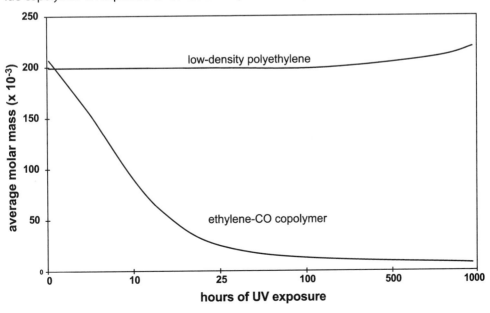

1995). Note that after only a few hours the copolymer has degraded to a fraction of its original molar mass, compared to the LDPE control. (The small increase in molar mass for LDPE is probably the result of crosslinking that competes with chain breaking.) Because the properties of polymers are in large part dependent upon molar mass, a plastic object made of the carbon monoxide-containing copolymer would rapidly become extremely weak and brittle if left in the sun. The rate of degradation would obviously differ from one location to another and depend upon the time of year. One would expect a much faster rate in Arizona in the summer than in upstate New York in December. The U. S. Congress passed a law requiring that, beginning in 1990, all ring carriers sold in the United States must be degradable. Photodegradable ethylene–carbon monoxide copolymers have commonly been used to satisfy this requirement.

Potentially, this same approach could be used to reduce other environmental litter, including some of the plastics that commonly end up as sea nurdles. For this approach to be totally successful, the polymer would need to progress through three distinct stages: use, physical degradation, and then biodegradation. The mechanical properties of the plastic objects need to remain stable throughout the expected period of use. Then, at the end of this period the physical properties need to decay rapidly. Finally, the initial breakdown products need to be biodegradable (see Figure 9-3). To date, photodegradable

Figure 9-3. The three stages of degradation (Scott 1995, with kind permission of Kluwer Academic Publishers).

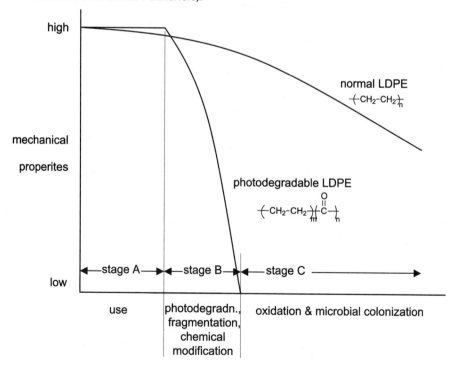

polymers have found application in only a few specialized areas, including agricultural mulching film and binder twine, in addition to beverage ring carriers. The use of mulching films produces higher yields of crops as well as increased quality. This, in addition to reduced costs for pesticides and weeding, more than pays for the increased cost of the copolymer. The twine used to bale straw and hay is usually plastic and can be a problem when it becomes entangled in farm implements. Photodegradable polyethylene-based fibers are commonly used for such applications.

Getting the Bugs Out—The Promise of Monomers from Nature

Consider an object made from an ethylene–carbon monoxide copolymer left sitting in a field or floating on the ocean. In a short time, photodegradation will cause it to weaken, and then wind, rain, and/or wave action will break it into very small pieces. However, as we discussed above, these pieces will themselves degrade only very slowly, because microorganisms will not attack them until they are oxidized and broken down into small-molecule metabolites. What would it take to make a polymer itself food for microbes? Such a polymer would be *biodegradable*—capable of being degraded by a biological system such as enzymes, bacteria, fungi, or algae. In fact, only a very few synthetic polymers are known that are truly biodegradable. They are step-growth polymers (i.e., are capable of being hydrolyzed), and contain monomer units that are natural products or are easily oxidized to natural products. Classes include certain polyesters, polycarbonates, and polyamides. Common examples include natural hydroxy acids, such as glycolic and lactic acids,

$$HO-CH_2-CO_2H \qquad\qquad CH_3-\underset{\underset{\displaystyle OH}{|}}{CH}-CO_2H$$

glycolic acid   lactic acid

as well as longer hydroxy acids such as 6-hydroxyhexanoic acid, which can readily cyclize to form a thermodynamically stable 7-membered lactone (caprolactone):

Equation 2

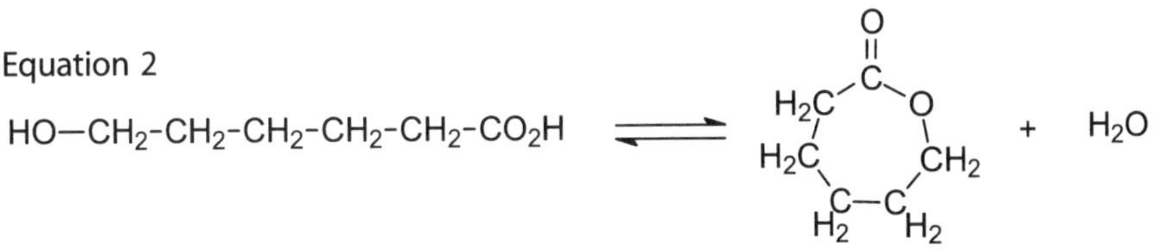

$$HO-CH_2\text{-}CH_2\text{-}CH_2\text{-}CH_2\text{-}CH_2\text{-}CO_2H \;\rightleftharpoons\;$$

$$+ \; H_2O$$

Polymers prepared from these kinds of monomers are desirable in principle for two reasons. First, the monomers come from *renewable sources* such as corn, wheat, rice, or even agricultural waste. (The fossil fuels petroleum, natural gas, and coal are considered nonrenewable sources of chemicals.) And second, because the monomers are natural products, the polymers and monomers can be metabolized by microorganisms (are biodegradable).

For much of the last century, scientists attempted to make useful plastics from hydroxy acids such as glycolic and lactic acids. Poly(glycolic acid) was first prepared in 1954, but was not commercially developed because of its poor thermal stability and ease of hydrolysis. It did not seem like a useful polymer. Approximately 20 years later it found use in medicine as the first synthetic suture material, useful *because* of its tendency to undergo hydrolysis. After the suture has served its function, the polymer biodegrades and the products are assimilated (Li and Vert 1995). Since then, suture materials, prosthetics, artificial skin, dental implants, and other surgical devices made from polymers and copolymers of hydroxy carboxylic acids have been commercialized (Edlund and Albertsson 2002).

Biodegradable polymers are useful in other areas of medicine. Consider, for example, drug delivery (Henry 2002). Taking a drug orally, say, that dissolves in the gut and is absorbed into the blood stream quickly produces a maximum concentration of the drug which then decreases in time. A second dose is administered hours later, and the cycle repeats. Peaks and valleys, peaks and valleys. For some drugs these concentration swings are not all that important. Some medications, however, provide optimum benefit over a relatively narrow concentration range. Below this, the drug may not be very effective. Above this, the drug may be toxic. See Figure 9-4.

Figure 9-4. Concentration of drug in bloodstream over time as a function of method of application (Santini et al. 2000, reprinted with permission of Wiley-VCH).

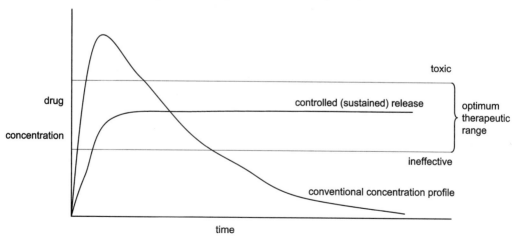

One way to obtain *sustained release* of a drug (continuous release to maintain near-constant concentration) is to mix it with a degradable polymer. The drug-polymer mixture can be administered as an oral tablet or implanted in the body. The active ingredient is constantly released as the polymer degrades. A medication for brain tumors called Gliadel takes advantage of this method. An antitumor drug (carmustine) is mixed with a polyanhydride and formed into a small wafer. The wafer is implanted in the brain, where the polymer is slowly hydrolyzed, releasing the drug. The drug is released at the site of the malignancy while the monomers from the polymer are metabolized (Sifton 2000).

Controlled release offers a number of advantages in addition to that of control of drug level. Drugs can be targeted to specific sites, reducing side effects and toxicity. In addition, the amount of drug administered is often less, as is the number of doses. Important advances in the area of pharmaceuticals will necessitate "intelligent delivery systems" (Langer 1998). These are devices using *microelectromechanical systems* (MEMS) microchips and other nanotechnology techniques to control precisely the timing and concentration of drug delivery (see Chapter 10). Professor Robert Langer's group at the Massachusetts Institute of Technology has published a recent review describing this technology (Santini et al. 2000). Encapsulation techniques such as those used for medications are also used to provide the controlled release of fertilizers and pesticides in agricultural products, flavors and vitamins in foods, and fragrances in cosmetics.

Other medical applications of biodegradable polymers include absorbable surgical implants, skin grafts, and bone fracture plates (Vert 2000). Some of these require higher strength materials than are possible with the aliphatic polyesters. Some approaches to increasing the strength but retaining biodegradability include synthesizing block copolymers, crosslinking, and fabricating composite materials (Huang and Edelman1995).

The Birth of a Bioplastic

Where do the monomers for these biodegradable polymers come from? Being natural products, they can be synthesized by natural organisms such as bacteria by fermenting sugars, organic acids, or alcohols. Many efforts to commercialize such products have been undertaken, with very few products yet to enter the marketplace. The primary problem is economic: Biodegradable copolymers have routinely been several times more expensive than large-volume polymers such as polyethylene and polypropylene. The first large-scale commercial synthetic polymer derived entirely from renewable resources such as grains has recently been introduced by Cargill Dow Polymers. Called NatureWorks, it is poly(lactic acid) or PLA, a polyester whose monomer comes from the fermentation of dextrose from corn. The company has built a new plant in Nebraska that is designed to produce 140,000 tonnes of PLA per year, using some 40,000 bushels of corn a day. The production of PLA requires less fossil fuel to make conventional petroleum-based plastics, thus reducing CO_2 emissions. Applications include fibers for clothing and clear packaging materials (Fahey 2001). Because the polymer is compostable, it can be used for food containers in fast-food restaurants in environmentally sensitive countries such as Austria and Sweden (Leaversuch 2002). Additional manufacturing plants are being planned in other parts of the world, using local feedstocks such as wheat, rice, or agricultural waste (Tullo 2000).

We have seen in this section that polymers can undergo several different types of degradation: thermal, mechanical, oxidative, chemical (e.g., hydrolysis), photo, and bio. For products to be useful, degradation needs to be carefully controlled. For most applications, resistance to any degradation is the safest approach. Sometimes, however, we want a material to retain mechanical strength for a given time and then rapidly degrade (e.g., agricultural mulch film). For other applications, we want a slow, controlled degradation with no induction period (e.g., sustained-release pharmaceuticals). For anything that might end up as litter, it would be very desirable that the polymer degrade within a reasonable time. Continued progress is being made in this direction. For example, lawn and leaf bags and some trash bags that are degradable are now being sold. The Eastman Chemical Company in the United States and BASF in Germany are both marketing a biodegradable polyester that can be used for food packaging as well as for trash bags. Their properties are similar to those of flexible PE film and can be processed on conventional equipment. Many communities in the United States now prohibit lawn and garden debris from being taken to sanitary landfills. *Composting* of this material is becoming more common. Therefore it makes sense to use degradable bags to collect the debris, and then compost the entire package. There is little reason to use degradable trash bags that are destined to end up in a landfill. Only those that become litter would actually degrade. In Europe, composting of household waste is much better established than it is in the United States, meaning that several companies are marketing biodegradable trash bags for use there.

Why don't more plastics fall into the category of degradable? Why don't more of the items we buy and dispose of degrade in the atmosphere? Three major reasons:

■ First, cost. Commodity polymers such as polyethylene, polypropylene, and polystyrene are synthesized on huge scales and are cheap. Historically, preparing copolymers (such as ethylene/CO) or substituting alternative polymers increased the cost substantially. As the demand for degradable plastics increases, the cost of substitutes will fall.

■ Second, control over the rate of degradation is necessary. Not all consumer items made of a photodegradable polymer would be stored or used under the same conditions. Some of them might begin to degrade prematurely, causing inconvenience or even safety concerns.

■ Finally, some manufacturers are concerned that degradable polymers cannot be reprocessed without a large decrease in mechanical properties. If this is the case, using degradable plastics would make recycling more difficult.

Composting of PLA cup over 47 days. Photos courtesy of Cargill-Dow, Inc.

Take the case of plastic fishing gear. Fishing line and nets are commonly made of monofilament nylon, an extremely strong, robust material as we have seen. A considerable amount of fishing gear is lost, either when it becomes tangled, caught on something, or simply discarded in the water (litter). Most of it floats, trapping a huge number of aquatic species and birds, as we pointed out earlier. A biodegradable substitute would have to possess two opposite properties. First, it would have to retain its strength for a long time (years?). Then, after it was no longer needed, or after it had become lost, it would have to degrade in the medium in which it was designed to be robust. Although it is difficult to imagine how to engineer a "trigger" into the material that would initiate the degradation when needed, this feat has been accomplished. Bernard "Bronco" Gordon, President of Polymer Chemistry Innovations, Inc., has invented and commercialized a fishing line made of Earthguard, a hydrodegradable plastic.

The neck of the center seal is entrapped in plastic fishing gear. Photo courtesy of National Oceanic and Atmospheric Administration, U.S. Department of Commerce.

Although commodity plastics is a mature industry, a huge amount of work remains to be done in the area of degradable polymers. New materials are being introduced at an ever-increasing rate. However, the opportunities for significant synthetic and technological advances, new polymers, and clever control mechanisms remain very high. This field should attract an increasing number of scientists with a mix of chemical, biological, materials science, engineering, and electronics backgrounds to address a growing array of challenging technological problems.

Polymer Recycling

Fortunately, we do not throw away everything. In a few states, consumers return soft drink and beer containers to stores to reclaim a deposit of 5 or 10 cents. And in many places, some types of plastics (in addition to glass bottles, metal cans, and paper and cardboard) are separated from the trash and *recycled*. What happens to these items? We want to understand why only certain plastics are recycled, how these are recycled, what some of the limitations are that limit their reuse, and what other options might exist to better utilize synthetic polymers.

Recycling is the process of collecting waste, sorting it, and reclaiming raw materials that can be sold to produce products. From our household waste, glass and metals have been recycled for decades in many parts of the United States. Automobiles have been extensively recycled in North America for many years. Approximately 75% of the mass of an average scrap car is recycled. Clearly this is a profitable business with an extensive infrastructure. First, a salvage dealer will remove any parts with resale value or components not suitable for shredding. The remainder then goes to a shredder, which reduces the car to fist-sized pieces. After magnets remove the ferrous metals, other materials are separated, some of which are recycled (e.g., metals such as aluminum, copper, zinc), leaving auto shredder residue (glass, sealers, sound deadeners, fabric, adhesives, paint, rubber, dirt, and plastics). Shredder residue amounts to 500 to 800 pounds per car, of which approximately 30% is plastic (Pett, Golovoy, and Labana 1995).

What about plastic consumer items? The recycling of plastic objects would seem like a relatively easy process—assuming that you could collect enough of them to make it worth your while. Assuming that you could identify the different types and then figure out how to separate them. And clean them and remove any labels. Assuming that you could figure out what to do with all of it once you had done all of the above. Unlike automobiles, for which recycling is driven primarily by the recovery of valuable steel, plastic objects are inherently low in value and are more difficult to identify and separate. Only some types are easily reprocessed.

First of all, it is important to keep in mind that recycling is a business. And the business is essentially that of selling our garbage to someone else. No one wants to buy our garbage unless there is something worthwhile in it. If it were really valuable, it would be stolen (Pearson 1992). Communities have organized collection and sorting operations for many years. However, companies will participate only if they see an opportunity to make money. This is true even in spite of (or perhaps because of) the involvement of state and local governments. Some regions require recycling programs as a part of municipal solid waste programs and some 11 states have instituted bottle deposit laws. Programs such as these provide a consistent supply of "raw materials," necessary if a company is going to commit money and resources for their recycling. In addition, some product or application must be identified from which a company can receive revenue (see Figure 9-5 and Table 9-1). Plastics processors have long recycled scrap from their manufacturing operations. This is called *primary recycling*. The industry built on collection and recycling of plastics from consumers (called *postconsumer* plastics) is relatively new.

What happens to a postconsumer plastic during recycling depends upon a number of factors. Several options exist, including reprocessing it as the same material (called *secondary recycling*), depolymerizing it to starting materials or heating (thermolyzing) it to produce small molecules (*tertiary recycling*), mixing it with other plastics, or using it for energy recovery (*quaternary recycling*). See Table 9-2.

Figure 9-5. The production, use, disposal of polymers, including recycling.

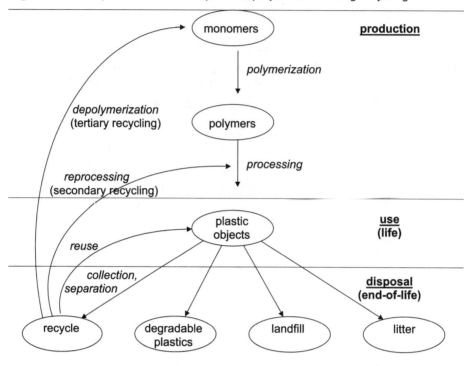

Table 9-1. Major constraints on the recycling of postconsumer plastics.

Successful recycling requires:	Recycling is not suitable for:
■ steady supply of large quantity of objects made of only one polymer	■ objects whose polymer makeup cannot be readily identified
■ consumer education, motivation, and establishment of recycling habits	■ objects that cannot be readily separated from other materials as pure polymer
■ existence of reliable collection infrastructure	■ polymers that when recycled would sell for more than virgin polymer
■ easy identification, separation, and cleaning	■ polymers that suffer substantial physical property degradation on reprocessing
■ steady market for recycled polymer	■ polymers that are contaminated with environmentally hazardous chemicals
■ opportunity to make a profit (commercial viability)	
■ technology based on scientific principles	
■ practical government regulations and public expectations	

Table 9-2. The major types of plastics recycling.

Type	Process	Product(s) Obtained	
primary (regrind)	reprocessing manufacturing scrap	same plastic object	polymers
secondary (mechanical)	reprocessing post-consumer polymers	different plastic objects	
tertiary (feedstock or chemical)	a. chemical depolymerization	monomer	small molecule chemicals
	b. thermolysis		
	hydrogenation	hydrocarbons	
	pyrolysis (no O_2)	gas, oil, C char	
	gasification (low O_2)	$CO + H_2$	
quaternary (energy recovery)	incineration (xs O_2)	$CO_2 + H_2O$ + heat	energy (heat, electricity)

Some Bright Spots: PET Bottles

Let's focus on PET as we discuss the major methods for recycling plastics. As we will see, this is one of the easiest polymers to recycle in large quantities. PET bottles are readily available and easily identifiable. Most plastic soft drink containers in North America, if not elsewhere, and an increasing number of juice, water and food bottles, are pure PET. We recognize them by the little "1" recycling symbol (and often "PET" or "PETE") on the bottom. They are collected in huge numbers, both through returns for deposits and from curbside recycling. Most are colorless, while a minority are green. They need to be separated from other plastics and from their caps (often polypropylene) and labels.

Hundreds of companies have been formed over the past 20 to 30 years to participate in plastics recycling, many of them dealing with PET. During that time, several hundred companies also went out of business, failing to find or maintain a profitable position in polymer recycling. During the 1990s, the relative costs for recycled and new (*virgin*) PET varied tremendously, with virgin resin being less expensive than recycled at times. The plastics recycling business, just like any recycling business, is extremely cyclical. This means that periodically the demand for recycled polymer is extremely low.

Get Fleeced

Can new uses be found for old materials? Sometimes opportunity is recognized only by the truly ingenious, and sometimes ingenuity comes from desperation.

In the early 1980s, before plastics recycling was well established, a New England businessman named Aaron Feuerstein had a problem. His company, Malden Mills, just about the last textile mill remaining in Massachusetts, desperately needed a successful product to avoid bankruptcy. By that time, most of the old textile mills in the northeast United States had become unprofitable and had closed, leaving thousands unemployed. Many companies had opened factories in the South where labor and taxes were considerably cheaper.

In experimenting with recycled PET, Feuerstein came up with a high-tech use for "inferior" polyester fiber. He invented a warm, synthetic fleece that he called Polar Fleece (later changed to Polartec). Then he struck a deal with the outfitting company Patagonia to market outdoor apparel made from his PET fleece (Fenichell 1996).

Synthetic fleece is lightweight, water resistant, and very warm. In addition, "fleece" garments became fashionable, in part, perhaps, because they were first marketed by an upscale retail company. By turning a cheap, readily available resource into a useful product, he saved his family mill and was able to continue to offer employment to local workers.

Reprocessing PET and Other Polymers: Secondary Recycling

Recycled PET can be easily separated, shredded, cleaned, and reprocessed into carpet fiber, insulation for sleeping bags, films for packaging, and nonfood bottles. This type of recycling is called *secondary recycling* (see Table 9-2). As we have mentioned previously, each time a polymer sample is processed, some thermal and oxidative degradation occurs, resulting in discoloration, decreased molar mass, and possibly increased crosslinking. Therefore, unless the recycled polymer is to be used for lower-value applications, recycled polymer is generally blended with virgin material during processing.

Other common thermoplastics can be similarly reprocessed, assuming that they can be identified, separated, and adequately cleaned. Bottles entering the recycling stream often contain caps made from a different kind of polymer, labels attached with hard-to-dissolve glue, foil seals, and residual product, among other impurities. Separating and cleaning the bottle polymer (say, HDPE) from these other materials can be time- and energy-consuming. Objects that contain only one kind of polymer and that can be readily identified and collected are much more readily recycled. For example, automobile batteries have been recycled for many years, primarily to recover the valuable lead metal from the electrodes. The outer case for almost all batteries is made only of polypropylene, and therefore these too are now usually recycled.

Not Everything Is Easily Recycled

Many plastic objects are not as readily recycled as PET or polyethylene bottles (HDPE). For one reason, identification and separation of different materials may be more difficult, increasing costs. One approach to recycling such materials is to reprocess mixed plastics for applications such as outdoor furniture, highway guardrails, decorative fencing, and plastic lumber. Processing different polymers together usually results in an *incompatible blend* in which the ultimate physical properties are relatively low. Thus such blends usually find use in "low-tech" applications such as those mentioned above. Although lumber made from recycled plastics continues to be relatively expensive, it is slowly finding applications, including use in locations near or under water (e.g., docks and piers) and in contact with soil. Restrictions on the heavy metals used to produce pressure-treated wood may open the way to using recycled comingled plastics as a replacement.

Back to Basics: Tertiary Recycling

What about food containers? Say we have collected a large number of perfectly good PET soft drink bottles. Why not just return them to the bottling plant and refill them just like we used to do with glass bottles? Plastic food packaging presents special challenges for recycling, because processors receiving used packaging materials have no control over what these materials were exposed to before entering the recycling stream. For example, someone may have temporarily stored a toxic pesticide in an empty PET soda bottle before recycling it. Traces of the pesticide could remain on or in the plastic even after vigorous cleaning, only to migrate from the plastic later after refilling, contaminating the product. Unlike glass, plastics can absorb some chemicals. For one thing, we know that they mix with small molecules called plasticizers, giving a mixture that has a lower T_g than that of the pure polymer. As we mentioned in Chapter 7, the migration of plasticizers to the surface of a plastic object can be a concern. If we can't clean and refill PET soft drink bottles, what can we do?

Recall that we discussed the reversibility of polyester formation in Chapters 4 and 5. Polyester producers have worked out a process called *methanolysis* in which shredded PET bottles are heated with methanol and catalysts, causing depolymerization of the polyester and producing dimethyl terephthalate (DMT) and ethylene glycol, the two starting materials from which the PET was prepared originally:

Equation 3

After the monomers are purified, they are used to produce new, *virgin* polymer. Any impurities present in the recycled PET are removed, either during methanolysis or during purification of the monomers. This process, sometimes referred to as *tertiary plastics recycling*, can be applied to other step-growth polymers such as nylon. Chain-growth polymers such as polyethylene and polypropylene do not so easily depolymerize to their original monomers. Heating them in the absence of oxygen (*thermolysis*) causes random carbon-carbon bond breakage, resulting in a mixture of small molecules. We will discuss thermolysis shortly.

So far we have considered the recycling only of thermoplastics. What about thermoset resins and other heavily crosslinked materials? Specifically, what about rubber? What can we do with all of those old automobile tires? Unfortunately, the characteristic that differentiates these materials from linear polymers also makes their recycling very difficult. By definition they cannot be remelted and reprocessed. So their recycling and reuse is severely limited. Throughout much of the twentieth century, new treads were applied to bald automobile tires, virtually doubling the life of the body of the tires. However, today only truck tires are routinely retreaded. Today's automobile tires are highly engineered composite materials containing different types of rubber in addition to polymeric fibers and steel reinforcing. The act of even shredding tires into small pieces requires substantial amounts of energy. Some thermoset resins are ground up and used as fillers in thermoplastic resins. Tires can be ground and used as filler in asphalt paving. Alternatively, they can be incinerated and the energy recovered (quaternary recycling) or thermally decomposed, as discussed below.

Recycling by Thermal Decomposition

We discussed the chemical depolymerization of step-growth polymers such as PET, which undergoes methanolysis to yield its original monomers. An additional form of tertiary recycling involves the thermal decomposition, or *thermolysis,* of polymers. If this is carried out in the absence of air, it is called *pyrolysis* and produces a mixture of organic compounds that can be refined just like petroleum. So this produces a cycle of petrochemical-polymer-petrochemical. If the thermal decomposition is carried out with a limited concentration of oxygen, the process is called *gasification*, and the primary products are CO and H_2. These gases can either be burned for their energy or separated and sold as chemicals. A third form of thermolysis is called *hydrogenation*, which produces a mixture of hydrocarbons not unlike gasoline or diesel fuel. See Table 9-2. A major advantage of thermolysis as a recycling strategy is that the starting materials can be a mixture of polymers and can even contain some level of nonpolymer contaminants. Therefore, sorting and cleaning are not generally necessary. In addition, thermoset materials are well suited to the technique.

Thermolysis differs from incineration (oxidation) in basic ways. *Incineration* (also called *quaternary recycling* or *waste-to-energy*) is complete combustion, with the major products being gases, primarily CO_2 and H_2O, plus heat. Thus incineration allows only the recovery of the *energy value* of the waste materials. In addition, there is public concern over the possibility of the formation of incomplete combustion products such as polynuclear organics (e.g., dioxins) in oxidative processes. Thermolysis, on the other hand, reclaims the *chemical value* of the waste polymer, producing a variety of compounds depending upon the specific method used. Thermolysis has been used with a variety of organic materials for many years. For example, *coal gasification* was heavily promoted in the 1970s and early 1980s as an alternative to petroleum as the source of small-molecule hydrocarbons.

Practical Recycling

We have covered the major methods by which polymers can be recycled. The question remains why so few polymers are actually recycled. We discussed some reasons above and hinted at a few others. So let's examine this question in a little more detail.

Only a few packaging polymers, those that are abundant and easily identified, can be recycled practically (e.g., PET beverage containers, HDPE bottles and jugs). Some polymers are part of a *closed-loop market*, and are therefore easily recycled. By this we mean products that are made of only one type of plastic and that are easily returned. Return-for-deposit beverage bottles are an excellent example. Also, bottled-water companies can recycle 5-gallon jugs from water coolers, because they are almost all made of the same polymer (polycarbonate) and the companies have to pick up the empties, anyway. Telephone companies often recycle telephones and other electronic equipment for the same reason. Carpet manufacturers cooperate with carpet retailers to recycle nylon carpeting, because many new carpet installations replace old, worn-out carpeting (Tullo 2000b). Single-use cameras are an interesting example. Often referred to (incorrectly) as disposable cameras, they are returned after exposure to a photo processor, who opens them and removes the film for developing. Most of the camera bodies are then returned to the manufacturer where the plastics and other parts are recycled into new cameras.

A related situation we'll call *piggyback markets*. By this we mean industries in which extensive recycling exists for some material other than polymers, but in which polymers are easily separated for recycle as well. We earlier mentioned the recycling of polypropylene battery cases. Infrastructure was already in place to recover valuable lead metal from the electrodes. It was easy to also recover the plastic cases when polymer recycling became feasible. Table 9-3 summarizes a few examples.

Table 9-3. Examples of "piggyback" recycling. The recycling of plastics in some products was added to existing infrastructure.

product	infrastructure for	bonus
automotive batteries	Pb electrodes	polypropylene cases
X-ray film	silver	PET film
electrical wire, cable	copper	LDPE and PP
automobiles and household appliances	steel, iron, copper	miscellaneous polymers

Many of the common polymeric materials that we encounter in our lives do not fall into the categories that we developed above. Some are not easily separated or identified, some are not easily cleaned, while others are easily identified but are not discarded in very large volumes. Table 9-4 categorizes some common polymers according to three criteria: separation/identification, volume generated, and whether they are part of a closed-loop collection. Although one could argue about some of the specific entries, the table is meant to help us better understand why only a few polymers are actively recycled.

Table 9-4. Comparison of the likelihood various plastic items will be recycled.

	Readily Identified Separated	High Volume?	Closed-Loop Collection?	Commonly Recycled?
Commodity plastics				
HDPE milk jugs	yes	yes	no	yes–household
PET juice bottles	yes	yes	maybe	yes–household
LDPE mustard containers	no	no	no	no
PS yogurt lids	no	no	no	no
PP sour cream containers	no	no	no	no
PP battery cases	yes	no	yes	yes–commercial
Engineering polymers				
nylon carpet fiber	yes	yes	yes	yes–commercial
polycarbonate sports H$_2$O bottles	no	no	no	no
20 L polycarbonate H$_2$O jugs	yes	no	yes	yes–commercial
automobile parts	some	some	yes	yes–commercial

No single method of recycling provides a practical method for handling polymer waste. To be effective, processes must be based on sound science and technology. In addition, a recycling program must make economic sense. There must be incentives for companies willing to participate to make a profit. Federal, state, and local government mandates need to take all of this into consideration, as well as accommodate for the differences from one geographical region to another. Transporting low-density polymer waste long distances for processing is not economically feasible. What works in metropolitan areas in the northeast United States would probably fail miserably in sparsely populated areas in the West. Further, not all polymers can be recycled practically. Therefore, recycling, degradation, landfilling, and energy recovery will no doubt all remain as important complementary disposal outlets for polymers.

Summing Up

After use, polymer products can take a variety of disposal paths. Many are simply discarded. Some of these end up as litter, while many are buried in landfills. Some are reprocessed and reused, while others are depolymerized or thermolyzed, affording monomer or other small organic molecules. Some are incinerated, producing energy. The vast majority of our petroleum is refined and burned as fuel, a practice that utilizes only its *heat energy*. A very small fraction of petroleum is converted into polymer, processed into useful objects, and then disposed of. A portion of this polymer is then burned as a fuel. For this portion, both the *chemical energy* and the heat energy are extracted from the original petroleum. This suggests that we should probably think of discarded polymer objects as a resource rather than simply as waste to be discarded.

Recycling is but one stage in the life cycle of a given polymer. We have become a throwaway society and are somewhat entrenched in habits that will make change difficult. If we truly need or want to lessen the amount of oil we require to support our lifestyles and to reduce the quantity of waste that enters landfills, we will need to adapt more of a "reuse" and "reduce" philosophy toward materials. Manufacturers make and sell what they think consumers want. We would need to change our pattern of consumption and encourage manufacturers to design products more for *reuse* rather than for *single use*. The trend worldwide is moving increasingly toward *sustainable development*, that is, development that meets the needs of the present generation without compromising the needs of future generations.

Polymers are becoming increasingly important in the area of controlled release, for drugs, flavors, fragrances, fertilizers, and pesticides, to name a few. Often the drug is suspended in a biodegradable polymer that is slowly degraded and metabolized (in other words, is a renewable resource).

Although most of our polymers come from petroleum, an increasing number are being prepared from natural monomers, a trend that will surely continue. The application of biotechnology to the synthesis of these monomers should lower their costs, making them and the polymers derived from them more competitive with petroleum-based products. Polymers prepared from natural monomers are usually biodegradable, meaning they can be degraded when placed in contact with soil or water and exposed to microorganisms (e.g., in compost piles or sewage treatment plants). Although these materials are particularly well suited for use as compostable lawn and leaf bags, they are finding more frequent application as nonpetroleum, renewable replacements for conventional plastics. Figure 9-6 compares the life cycle of an LDPE film with that of one prepared from lactic acid. Note the basic differences in two of the steps in the cycle. Ethylene comes from the chemical reactions that take place in the refining of petroleum. Lactic acid

Figure 9-6. Comparison of the life cycles for a polymer produced from petroleum (LDPE) with one produced from corn (PLA).

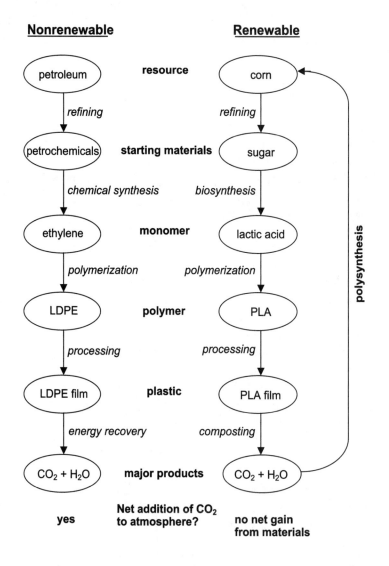

for the Cargill-Dow process (see "The Birth of a Bioplastic" earlier in this chapter) is *biosynthesized* by *lactobacilli* bacteria, the same species used to make yogurt. The laboratory synthesis of lactic acid is considerably more expensive than the biosynthesis. Also, there is no net increase in atmospheric CO_2 on the degradation of PLA (based on material), because the original resource (corn) removed CO_2 from the atmosphere to produce sugar. The manufacturing steps, however, require considerable energy from nonrenewable sources.

References Cited

Edlund, U., and A.C. Albertsson. 2002. Degradable polymer microspheres for controlled drug delivery. *Advances in Polymer Science* 157: 67–112.

Fahey, J. 2001. Shucking Petroleum. *Forbes* 168 (13): 206–08 (November 26).

Fenichell, S. 1996. *Plastic: The making of a synthetic century*. New York, NY: HarperBusiness.

Harlan, G., and C. Kmiec. 1995. Ethylene-carbon monoxide copolymers. In *Degradable polymers: Principles and applications*, eds. G. Scott and D. Gilead, 159. New York, NY: Chapman and Hall.

Henry, C. M. 2002. Drug Delivery. *Chemical and Engineering News* August 26: 39–47.

Huang, S. J., and P. G. Edelman. 1995. An overview of biodegradable polymers and biodegradation of polymers. In *Degradable polymers: Principles and applications*, eds. G. Scott and D. Gilead, ch 2. New York, NY: Chapman and Hall.

Krajick, K. 2001. Message in a bottle. *Smithsonian* 32(4) (July): 36–47.

Langer, R. 1998. Drug delivery and targeting. *Nature* 392 (Suppl.): 5–10.

Leaversuch, R. 2002. Renewable PLA polymer gets 'green light' for packaging uses. *Plastics Technology* 48 (3) (March): 50–55.

Li, S., and M. Vert. 1995. Biodegradation of aliphatic polyesters. In *Degradable polymers: Principles and applications*, eds. G. Scott and D. Gilead, 44. New York, NY: Chapman and Hall.

Pearson, W. 1992. Recycling plastics from municipal solid waste: An overview. In *Emerging technologies in plastics recycling*, eds. G. D. Andrews and P. M. Subramanian, 1. ACS Symposium Series 513. Washington, DC: American Chemical Society.

Pett, R. A., A. Golovoy, and S. S. Labana. 1995. Automotive recycling. In *Plastics, rubber, and paper recycling: A pragmatic approach*, eds. C. P. Rader, S. D. Baldwin, D. D. Cornell, G. D. Sadler, and R. F. Stockel, 49. ACS Symposium Series 609. Washington, DC: American Chemical Society.

Rader, C. P. 1995. Polymer recycling: An overview. In *Plastics, rubber, and paper recycling: A pragmatic approach*, eds. C. P. Rader, S. D. Baldwin, D. D. Cornell, G. D. Sadler, and R. F. Stockel, 2. ACS Symposium Series 609. Washington, DC: American Chemical Society.

Rathje, W. L. 1989. Rubbish! *The Atlantic Monthly* 264(6) (December): 99–109.

Santini, J. T., A. C. Richards, R. Scheidt, M. J. Cima, and R. Langer. 2000. Microchips as controlled drug-delivery devices. *Angewandte Chemie International Edition in English* 39: 2396–2407.

Scott, G. 1995. Photo-biodegradable plastics. In *Degradable polymers: Principles and applications*, eds. G. Scott and D. Gilead, 170. New York, NY: Chapman and Hall.

Sifton, D. W, ed. 2000. *Physicians' desk reference*, 54th ed., 2556-57. Montvale, NJ: Medical Economics Co.

Tullo, A. 2000. Plastic found at the end of the maize. *Chemical and Engineering News*, January 17: 13.

Tullo, A. 2000. DuPont, Evergreen to recycle carpet forever. *Chemical and Engineering News*, January 24: 23–24.

Vert, M. 2000. Lactide polymerization faced with therapeutic application requirements. *Macromolecular Symposia* 153: 333–42.

Wool, R. P. 1995. The science and engineering of polymer composite degradation. In *Degradable polymers: Principles and applications*, eds. G. Scott and D. Gilead, 141. New York, NY: Chapman and Hall.

Other Reading

Grosberg, A. Y., and A. R. Khokhlov, 1997. *Giant molecules: Here, there, and everywhere*. New York, NY: Academic Press.

Gross, R. A., and B. Kalra. 2002. Biodegradable polymers for the environment. *Science* 297: 803–07.

Rader, C. P., S. D. Baldwin, D. D. Cornell, G. D. Sadler, and R. F. Stockel, eds. 1995. *Plastics, rubber, and paper recycling: A pragmatic approach*. ACS Symposium Series 609. Washington, DC: American Chemical Society.

Chapter *10*

A Glimpse of Things to Come

No one can accurately predict the future, not even meteorologists. Reading older literature that confidently predicted polymer sales volumes or revenues in x numbers of years is usually pretty amusing. In this chapter we will not attempt to look very far into the future. However, as we conclude this introduction to a field of physical science that essentially evolved over the duration of the twentieth century, it might be instructive to reflect a little on the current state of polymer chemistry. Based upon what is now on the drawing board, we can probably make some pretty good guesses as to what to expect for the next few years.

How Did We Get Here? Evolution and Revolution

Scientific and technological progress, like other areas of human endeavor, advances by a cyclical process of gradual *evolution* interspersed with occasions of scientific *revolution* (Mark 1987). For example, let's consider Figure 10-1, which illustrates the development of polymer science through the commercialization of one family of polymers, polyolefins (Sinclair 2001). One could construct a similar plot for other families of polymers. Revolutionary episodes

Figure 10-1. Evolution and revolution in the development of polyolefins (adapted from Sinclair 2001 and reprinted with permission of Wiley-VCH).

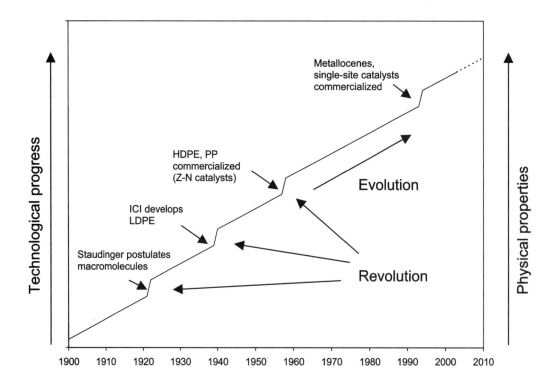

include intellectual contributions, such as Staudinger's convincing arguments that polymers are made up of very large molecules. Revolution also includes radical technological breakthroughs, such as the development of new classes of polymers or of catalysts that significantly change the properties of a given class of materials.

In the specific case of polyolefins, this cycle of evolution and revolution is remarkable in that it has produced both new families of polymers as well as drastic improvements in the properties of older ones. Much cheaper commodity polyolefins can now replace steel, wood, and expensive engineering polymers in many applications. For example, consider a part, such as a cargo load floor, for the trunk of an automobile or SUV. Until recently, this part would typically have been fabricated from a variety of materials, perhaps particleboard glued to a piece of carpeting with steel brackets and a steel hinge attached. Today the same part is probably made of only one polymer, polypropylene (PP). This is possible because a polymer, as we discussed in Chapter 8, can have different physical properties depending upon how it is processed. The core has a honeycombed cellular structure for strength and is sandwiched between two skin layers. Decorative carpeting, also made of polypropylene, can be attached to one side. The part can be mass-produced very quickly, including a PP "living" hinge that can be flexed innumerable times without breaking. The polypropylene part has half the mass and is produced at twice the rate of an older cargo load floor. And, unlike parts made with conventional composites that contain a mixture of polymers and other materials, the all-polypropylene part can be very easily recycled (APME 2001). This theme, coupling the availability of new materials with innovation in processing technology, continues to be repeated throughout the manufacturing industry, showing up in many of the new products that enter the market every day.

A View of the Future: From Inert Polymers to Active Materials to Smart Systems

"Active" Materials

Many of the applications we have discussed so far use polymers in only an inert or passive role. They serve as objects or as containers or as films that cover something. The choice of the specific polymer depends upon the application. As we have seen, some polymers resist solvents better than others and some are better barriers to particular gases than others. Traditionally, a specific polymer was chosen for a specific application, say a food-packaging application, based on a number of variables, including strength, appearance, barrier properties, compatibility with the product, and cost. For perishable foods such as produce, inexpensive plastic packaging has helped make fruits and vegetables widely available year-round in many societies. We have mentioned that, without efficient packaging, the amount of food lost to spoilage would be enormous. The barrier properties of a given polymer depend upon the structure of the polymer and its thickness—in other words, on the physical properties of the polymer making up the packaging material.

But maintaining quality involves more than just protecting food from the outside environment. In fact, for many years the food industry has modified the atmosphere inside food containers to retard spoilage. Many foods maintain their quality much longer when oxygen is replaced with CO_2 or nitrogen. These gases help reduce the growth of microbes and reduce the rate of chemical reactions in the food. For this to work, the packaging material must prevent these gases from escaping from the container and prevent the influx of oxygen from the atmosphere. Using this concept is effective only for certain foods, however. Products such as bread tend to release oxygen slowly from internal pores, causing the oxygen level to increase over time.

Other foods, such as fresh fruits and vegetables, consume oxygen during storage and release CO_2. To maintain freshness, packages need to allow CO_2 to diffuse out and O_2 to diffuse in. Keeping the oxygen concentration low slows the respiration rate and keeps the food fresh longer, but if the O_2 concentration falls too low, the food quickly decays anaerobically. Plastics are ideal for this application because the barrier property of the packaging can be controlled by the choice of material(s) and its thickness.

What if we could produce *active packaging*, a polymeric container or packaging film that was designed to *control* concentrations of gases inside the package? Using such a technique should protect perishable food for even longer periods, or allow packaging and shipping that was previously impractical. In this technology, additives are incorporated in the packaging material that absorb gases such as oxygen. Sometimes the active ingredient is placed only in a bottle cap or in a label rather than throughout the entire package. Besides oxygen, active materials absorb ethylene, ethyl acetate, ethanol, ammonia, and hydrogen sulfide, undesirable gases that form inside packages of fruits, vegetables, and other foods and hasten their decomposition (Graff and Moore 1998). The use of such packaging has changed the way some produce is sold. Prewashed, ready-to-eat salads and peeled carrots are now available that can be kept fresh for two weeks or more without wilting or spoiling. Other types of active packaging involve the control of moisture, important in dried food products, or the release of low concentrations of ethanol to retard mold and yeast growth. Active packaging is an increasingly important approach for the protection of foods against spoilage. For one thing, it can reduce or eliminate the need to add chemical additives directly to food.

"Smart" Materials

The term *smart materials* is defined differently by different people. Essentially it refers to materials that sense some kind of stress or environmental change and respond in an appropriate manner. Basically they are active materials with a *sensor*. Smart packaging has all of the attributes discussed above under active packaging but reveals information about the quality of the product contained inside. For perishable foods, this might be a patch that changes color if the package has been exposed to too high a temperature or has reached the end of its shelf life. A sensor in a fish package might react with volatile amines and change colors, indicating that the fish was no longer fresh.

Materials with a "Memory"

Polymers (or other materials) are said to have a *memory* if upon some treatment, they revert to an initial shape or condition. They are called *shape-memory polymers*. In simplest form, they are illustrated by a stretched rubber band or a deformed metal spring, because both of them will relax to their original shapes when the stress is removed. In other words, even though they have been deformed, they "remember" how they started. We can design a polymer to do this by fabricating it in a desired shape and then "fixing" it by crosslinking. Heating it above its T_g, deforming it to a new shape, and then cooling below T_g will give the polymer a new, temporary shape, although the sample will be under stress. When it is reheated above T_g, it will revert to its original shape to relieve the stress. Physical crosslinking rather than chemical crosslinking can also be used to freeze in a structure. Examples of common materials that exhibit a shape-memory effect are shrink-wrap film and heat-shrink tubing. Shrink-wrap film is useful because we can loosely wrap an object or cover a window and then heat the film to cause it to return to its original dimensions and form a tight seal. PET (polyethylene terephthalate) soda bottles are another example of an object exhibiting shape memory, although this property is not taken advantage of in the bottle. We discussed the stretching of PET bottles in the section on blow molding in Chapter 8.

How can we take advantage of this property in a more sophisticated application? One of the objectives of modern surgery is to minimize invasive techniques such as by making smaller incisions. For many operations, such as arthroscopic knee reconstruction, this approach has been extremely successful. However, tiny incisions make the job of tying small knots with sutures very difficult and the likelihood of implanting large devices almost impossible. Robert Langer and colleagues (Lendlein and Langer 2002) are developing biodegradable polymeric devices that revert to their original shape when being warmed to body temperature. Thus, a bulky device could conceivably be temporarily fabricated in a compact shape, inserted cool through a small incision, whereupon it would expand to its desired size and shape when it reached body temperature. Likewise, sutures can be prepared from a fiber that has a T_g very close to body temperature. Stretching the fiber above this temperature and then cooling it freezes in the temporary structure.

Langer has shown that if this material is inserted loosely in an incision and then warmed, the suture pulls the incision closed as the fiber shrinks to its original length.

Smart materials need a sensor mechanism of some kind, such as chemical (e.g., color change), physical (e.g., shape-memory), piezoelectric, magnetostrictive, or optical fiber. A *piezoelectric* device converts electrical energy into mechanical energy, or vice versa. Lighters for outdoor grills or gas stoves often use piezoelectric devices to ignite the gas. A *magnetostrictive* material converts magnetic energy into mechanical energy and vice versa. *Optical fibers* use light energy to measure strains, temperature, pressure, or electrical or magnetic fields.

Over time, smart materials will become increasingly common in a wide variety of products (Mestel 2002). Some garments are already available with fiber-optic sensors to monitor heart rate, breathing rate, and temperature, among other things. We will probably see shirts, socks, or other articles of clothing that can destroy sweat and odor-causing chemicals, thereby reducing the need for cleaning. Some individuals might benefit from a jacket that self-inflates when its wearer begins to fall, protecting against injury much like an automobile air bag. The military is developing clothing that can change color like a chameleon, detect gases, signal location, and dispense medications if necessary.

Smart technology is being used with increasing frequency and sophistication in more complex systems. Not surprisingly, many developments occur for military applications. Modern aircraft, for example, use smart technology to monitor and reduce vibrations and noise, to reduce drag and turbulence, and to detect impact. Toyota is adapting vibration-reducing technology to automobiles. Smart shock absorbers will detect and analyze road vibrations and adjust the shock absorber to eliminate the vibration from the automobile (Newnham and Amin 1999). The windows of modern buildings will be self-adjusting to control the amount of sunlight and heat able to pass through. Medical devices are becoming smaller and increasingly sophisticated, measuring concentrations and dispensing medications as needed. Consumer products, too, are being affected. Good examples can be found in high-end sporting equipment such as tennis rackets, baseball bats, golf clubs, snowboards, and skis (Akhras 2000). The primary objective in these applications is the reduction of vibration, thus increasing comfort, reducing injury, and maximizing performance.

Smart materials such as these illustrate an important technological direction for materials science: the design of materials with sophisticated properties that behave more like biological systems. Let's briefly recap our history. In Chapter 4 we noted a significant period of discovery when people modified natural polymers to improve their properties. We can call this period, roughly before 1900, Stage 1, and it asked the question, "How can I improve upon nature?" This was followed by a century of synthetic polymer science in which non-natural monomers were converted by a variety of chemical reactions into polymers that have little resemblance to those that nature provides. This was the Plastics Age, the era of commodity polymers, engineering plastics, and high-performance materials. We'll call the synthetic age Stage 2, essentially spanning the twentieth century and asking the question, "How can I make a synthetic material that's better/cheaper/stronger than a natural polymer or a polymer derived from nature?" Stage 3, the stage we are now in, is asking a different set of questions. "How can we make polymers that approach the sophistication of natural polymers such as proteins and nucleic acids? How can we make them as efficiently as natural polymers are made (in water at ambient temperatures and pressures) and so that they self-assemble into useful structures (no expensive fabrication needed) and degrade naturally (biodegrade) when their useful life is over?" We are entering a period where we are trying to more closely learn from and adapt nature's ways, a process called *biomimicry*. The scientific and technical challenges in this stage are both enormous and fascinating. To solve them will require the abilities of a large number of creative scientists, people whose scientific understanding transcends the traditional boundaries of chemistry, physics, mathematics, and biology.

Thinking Smaller and Smaller: Macro, Micro … Nano

We have all lived through a time of miniaturization of electronic devices. Computers, TV sets, audio equipment, cellular telephones, and, obviously, the electronic components inside them, continue to shrink in size. Nothing dates a movie from the 1990s quite as much as seeing someone holding an old cell phone. One of the factors that allows smaller consumer products is thinner yet stronger materials.

Another factor that contributes to much of this miniaturization is smaller and smaller microelectronic components, allowing huge increases in the number of devices on one chip. As we mentioned in Chapter 4, today's computer chips contain more than 30 million transistors. Similar science and technology are being applied to make other materials and useful devices containing extremely small components.

Materials on the Nano Scale

Many materials exist that have dimensions in the range of 1 nm to several micrometers. Recall that colloidal particles (e.g., latex particles from emulsion polymerization, colloidal silica or alumina, etc.) fall in the range from about 10 nm to 1000 nm (1 μm). A few examples of nanoparticles that are designed with more specific structures or geometries include carbon nanotubes, metal clusters, nanoscale magnetic crystals, and semiconducting nanocrystals ("quantum dots"). When some of these particles are incorporated into nanocomposites, entirely new technologies emerge that promise to affect our lives for years to come.

Nanocomposites

Recall that in Chapter 7 we discussed the idea of reinforcing polymers by making composites. Particles (spherical or other shapes) or fibers are included during processing to form materials that have strength and properties superior to either single constituent part. Such composites can be called macro- or microcomposites, meaning that the scale of the two phases, the discontinuous additive phase and the continuous polymer phase, are often large enough to be seen by the naked eye or under a microscope. As we mentioned, for the composite to have superior strength, the adhesion between the different phases must be very high. Oftentimes the particles making up the additive must be treated in some way to improve this adhesion, a process that is not always inexpensive or totally effective. Inorganic additives or reinforcing agents, which are typical, have relatively high densities, potentially increasing the mass of a fabricated part.

So what happens if we make the additive particle size smaller and smaller? For much of the 1990s, many scientists researched *nanocomposites*, multiphase systems whose constituent parts have dimensions on the order of nanometers. Often the additives are inorganic *nanoparticles*, and can be spherical, one-dimensional rods such as carbon nanotubes or Al_2O_3 microfibers (nanometer-scale diameters and micrometer-scale lengths), or two-dimensional, layered structures such as clays (nanometers in thickness and micrometers in area). Because these particles are so small, their surface character dictates the properties they contribute to the nanocomposite. To gain an appreciation of this, consider slicing a cube 1 micrometer (1000 nm) on a side into thinner and thinner pieces. The volume of each slice decreases in direct proportion to the decrease in thickness. The surface area, however, decreases proportionately much less because it depends primarily on the combined area of the top and bottom surfaces (which does not change). We can state this more simply by saying that, as the particles become smaller, the surface-to-volume ratio increases. This is illustrated in Figure 10-2, which plots the surface area/unit volume as a function of thickness. Note that both axes are log scales. In addition, because nanoparticles have such high surface areas, often only 3 to 5 mass percent provides optimum effect. In other words, a very small amount of nanoparticle produces very large improvements in physical properties for the nanocomposite. This provides a huge advantage in many applications, because the density of the composite is not much greater than that of the pure polymer. Therefore, fabricated parts weigh considerably less than those made of conventional composites.

Scientists are finding that the incorporation of such small particles in polymers changes the behavior of polymers in unexpected ways. For example, they enable polymer chains to organize differently, often leading to increased crystallinity. The interaction of extremely small particles with light changes as a function of their size. In other words, varying the size of the nanoparticle

Figure 10-2. The effect of thickness on the surface-to-volume ratio for a rectangular nanoparticle. Consider slicing a cube 1 micrometer on a side, producing thinner and thinner pieces.

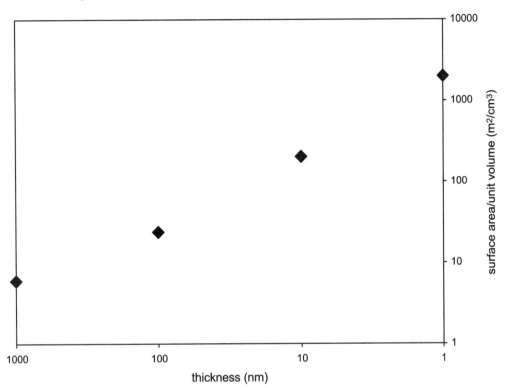

changes its color. Therefore, one can obtain brightly colored materials without having to add separate pigments or dyes. These small particles scatter light in interesting ways also, suggesting the possibility of constructing nanometer-sized lasers and other unique and potentially useful optical devices.

Nanocomposites often have improved physical properties with few if any negative effects. This is in contrast to conventional composites, which usually trade off one property to gain another. For example, a conventional inorganic reinforcing agent put in a polymer to produce a higher modulus often causes a reduction in elongation and almost always causes an increase in density. Considerable work focuses on attempts to improve the physical properties of biodegradable polymers by making nanocomposites. One area that might benefit from these efforts is drug delivery. Table 10-1 summarizes some of the properties that nanocomposites can potentially affect, as well as examples of possible end-use applications. Some have suggested that this is a method that essentially makes new materials without the need to synthesize new polymers (Giannelis 2002). This is a very significant point, because only a few companies produce almost all of our polymers, while a very large

Table 10-1. Properties that can be improved with nanocomposites and some potential applications (courtesy of Prof. Emmanuel P. Giannelis, Cornell University).

Property	Application
heat stability	automobile engine parts
gas permeability	food packaging (e.g., beer bottles)
impact resistance	automobile bodies
scratch resistance	decorative coatings
flame resistance	airplane interiors
biodegradation	tissue engineering, drug delivery
charge generation	photovoltaic devices
ionic conductivity	batteries
temperature or pH reversibility	microfluidic devices

number of compounders and fabricators produce the many products made from them.

Synthetic polymers are not the only macromolecules that interact with nanoparticles. Examples of *natural nanocomposites* abound and include bones, teeth, cartilage, scales, and shells. They consist of inorganic minerals such as carbonates and phosphates in combination with biopolymers such as carbohydrates and proteins. The possibility of mimicking these natural materials has captured the interest of biomedical researchers. For example, it is possible to encapsulate functionalized nanoparticles with natural polymers and deliver the nanocomposite to a specific site such as cancerous tissue. Once there, the nanoparticle could be activated and perform some therapeutic function. For example, it might absorb near-infrared energy and become warm. The localized heat could be used to kill malignant cells. Or the heat could cause a change in the properties of the nanoparticle, releasing a potent drug precisely at the diseased site. Other potential uses include the placement of nanocomposites in surgical sites to improve adhesion or to provide scaffolding for bone and tissue growth. For an overview of the field of polymer nanocomposites, see "Polymer Nanocomposition Approach to Advanced Materials" (Oriakhi 2000). "Less Is More in Medicine" reviews nanotechnology as it affects biomedical research and medicine (Alivisatos 2001).

Smaller and Smaller Devices

To make some nanosized materials even more useful and smarter, we need to give them moving parts. We need nanodevices with gears, levers, and motors. We need tiny pipes, pumps, and valves to transport liquids. This would enable the generation of a whole new scale of nanomachines and nanosensors.

One outcome of the effort to make smaller and smaller devices is called *microelectromechanical systems* (MEMS), or sometimes *micromachines*. This is analogous to the concept of integrated circuits in that a complex array of mechanical components is fabricated on a silicon wafer or synthetic polymer. For example, a series of extremely small mirrors, connected by gears and linkages to micromotors, could be used as switches in optical fiber networks, routing signals to the correct location. This would avoid the necessity of converting light signals to electrical signals at each exchange. Figure 10-3 shows a scanning electron micrograph of part of a small MEMS gear assembly, alongside red blood cells and a pollen grain to provide an appreciation of the scale. The bar under "50 μ" measures 50 micrometers (50,000 nm). Such machines are fabricated using microlithography, analogous to photolithography but using electron beams rather than visible or ultraviolet light for exposure. This is quite a slow process, not yet suitable for the mass production of large numbers of devices.

Microfluidics, the flow of minute streams of fluids through channels tens of micrometers wide, is another technology that is benefiting from microlithographic techniques. For example, a complex device might have 1000 chambers connected by a series of microchannels and controlled by some 3500 valves, all individually addressable and contained on a 1 inch x 1 inch PDMS [polydimethylsiloxane]

Figure 10-3. Scanning electron micrograph of microgear and lever alongside red blood cells and pollen grain (courtesy of Sandia National Laboratories 2003). Reference bar is 50 micrometers in length.

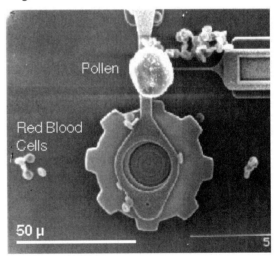

chip (Henry 2002). Such devices will essentially function as a *lab-on-a-chip*, useful for very rapid, automated chemical and biochemical assays such as genetic analysis, environmental microbiology, and drug screening. The technology might make possible easy and reliable home blood tests using only a pinprick-sized sample. Conceivably, the chip could be connected to a home computer and the results e-mailed to a physician for diagnosis (Knight 2002).

Electronic paper has been hyped as the technology that will replace products printed on conventional paper such as books and newspapers. Combining the features of both paper and computer displays, e-paper will be lightweight, flexible, and easy to read from many angles in all kinds of light. Unlike traditional printed products, the information displayed on e-paper would not be permanent, however, but could be changed, even animated. A newspaper, for example, could have a given story updated during the day as new events unfold. Other potential applications include wallpaper whose images can be changed rapidly, or advertising or in-store displays in which the messages can be changed remotely. In grocery stores, for example, workers would no longer need to walk up and down the aisles changing the prices of each item on each individual shelf. Signals would be sent out from a centralized computer instead. High-resolution devices will require both sophisticated polymeric materials and clever device engineering to produce practical products.

From Macromolecules to the Supramolecular

Just as integrated circuitry completely revolutionized electronics, supramolecular science promises to provide us with an exciting array of new technologies. *Supramolecular chemistry* means chemistry beyond the molecule. It utilizes *nanotechnology*, the assembly of complex arrays of components that are held together by intermolecular forces rather than by chemical bonds. Supramolecular chemistry requires the placement of atoms and molecules in specific locations in three-dimensional structures considerably larger than most molecules. We might consider this a *bottom-up* approach, meaning that we construct the final object from a selection of small pieces. But how can this be done?

A nanostructure has dimensions on the order of 1 to less than 100 nanometers. A solid-state physicist or electrical engineer used to designing integrated circuits using a *top-down* approach would consider this quite small. Currently, 90 nm is considered by many to be the practical lower limit for features mass-produced by photolithography. Building devices, electronic or otherwise, with considerably smaller features requires different technology. Although techniques do exist for fabricating devices with features smaller than 90 nm, they tend to be quite slow. We should appreciate that the top-down approach has been used successfully for much of the twentieth century to provide a wide range of devices and materials.

Unlike the physicist or electrical engineer, a chemist would consider a nanostructure quite large. Whitesides estimates that such a material would consist of 10^3 to 10^9 atoms and have a molar mass of 10^4 to 10^{10} (Whitesides, Mathias, and Seto 1991).

To the biologist, however, 1 to 100 nm is familiar territory, because a number of natural structures fall in this range. Figure 10-4 compares a few biological, synthetic chemical, and fabricated objects that range in size from approximately 0.1 to 100,000 nm (100 micrometers). Comparable regions of the electromagnetic spectrum appear at the right for comparison. Note that the length scale is logarithmic.

Figure 10-4. Relative sizes of biological and synthetic objects, with a comparison to comparable regions of the electromagnetic spectrum (adapted from Whitesides 1991).

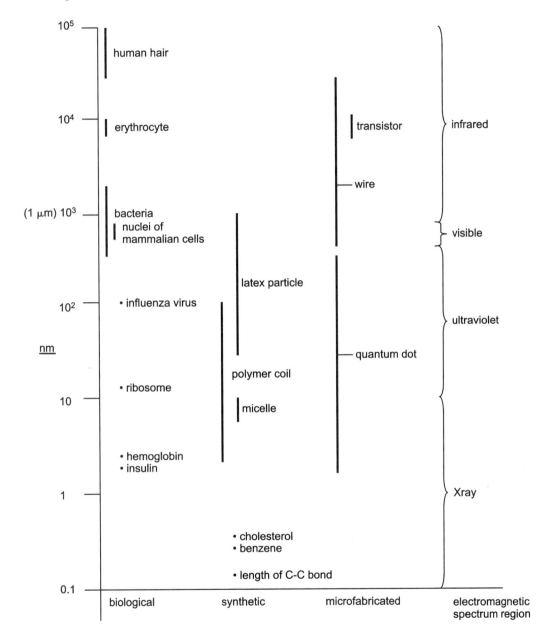

Approaching the problem from the *bottom up*, the chemist would be seriously challenged to design techniques for the synthesis of nanosized objects with specific geometries and function. Although synthetic methods have become much more sophisticated, the precise control of the three-dimensional structure of macromolecules, let alone assemblies of macromolecules, is in its infancy. Perhaps it is time to take a few lessons from nature.

The Concept of Self-Assembly

Self-assembly is the construction of molecular structures in a single step using noncovalent forces of attraction (MacGillivray and Atwood 2001). Although natural polymers such as proteins have an astronomical number of possible conformations available, by some mechanism they form specific, well-defined structures based on only one conformation. For proteins, the ultimate structure is determined by the sequence of amino acids that are covalently bonded in their chains. This sequence favors specific conformations that give rise to higher orders of structure. These structures are held together by specific intermolecular interactions using secondary bonding forces (e.g., hydrogen bonding, dipole-dipole, and van der Waals forces). Biological synthesis often involves the use of templates to generate large, complex three-dimensional structures.

Synthetically, self-assembly has been pretty much limited to very specific circumstances. One example is the generation of monolayers or bilayers on specially prepared surfaces. This technique requires special equipment to ensure that only a single layer of molecules is laid down at a time. An example is shown in Figure 10-5, which depicts a gold metal surface covered with an array of small molecules. Each molecule consists of a long hydrocarbon chain with a hydrophilic sodium carboxylate group (CO_2^- Na^+) at one end. The molecules are bonded to the gold surface at the hydrophobic hydrocarbon end. Under normal circumstances, the hydrocarbon chains extend away from the gold surface and stand parallel to each other much like blades of grass in a lawn. The outer surface is covered with the hydrophilic groups. Interestingly, the layer is actually switchable. When an electrical potential is applied to the gold underlayer, the negatively charged carboxylate groups are attracted to it, bending over and exposing the hydrophobic alkane chains. This changes the nature of the outer surface of the monolayer from hydrophilic to hydrophobic (Lahann et al. 2003). Potential applications for a switchable surface such as this include data storage and the microfabrication of controlled-release devices.

Another class of materials that tends to self-assemble is *liquid crystals* (LCs). As their name implies, liquid crystals are a state of matter somewhat between the disorder of liquids and the order of crystals. The molecules diffuse randomly as do liquids, yet maintain some order characteristic of solids. Liquid crystals are commonly used in displays such as digital watches, calcula-

Figure 10-5. Functionalized surface that switches from hydrophilic to hydrophobic when electric charge is applied (Lahann et al. 2003, courtesy of Prof. Robert Langor).

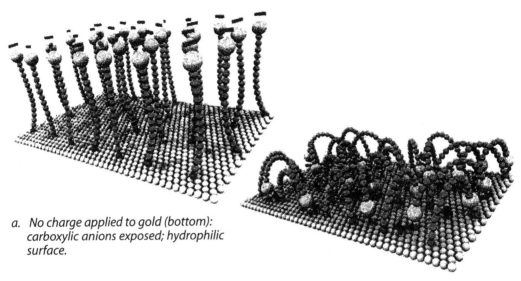

a. *No charge applied to gold (bottom): carboxylic anions exposed; hydrophilic surface.*

b. *Positive charge applied to gold, attracting carboxylic anions: hydrocarbon chains exposed; hydrophobic surface.*

tors, audio equipment, cellular telephones, and some flat-panel displays such as laptop computers. Newer applications include large devices such as curtainless privacy windows, which can be converted from transparent to translucent at the flip of a switch. LCs are often small molecules but they can be polymeric. The polyaramid Kevlar is a common example of a polymeric LC. Its solutions are in the liquid crystalline state. When spun from solution as a fiber, the rigid molecules align themselves and form a network of intermolecular hydrogen bonds, resulting in a very strong material.

A key challenge is figuring out how to get less exotic polymers to self-assemble. Physical scientists are a long way from being able to control polymer syntheses so that the self-assembly of complicated, three-dimensional molecules is possible. This is a very active area of research, one that promises eventually to result in interesting new scientific understanding, new technological tools, and new materials with unique properties.

Until recently, electronic devices have been based to a large extent on semiconducting inorganic materials. Integrated circuits developed around silicon technology, while light-emitting diodes have been based on gallium arsenide materials. For a variety of reasons, organic materials, including synthetic polymers, are becoming increasingly important. Some of the reasons include a greater ease of processing and the ability to make flexible, lighter devices, potentially at reduced costs. *Organic light-emitting diodes* (OLEDs) are now found in some commercial applications such as digital camera displays. Their polymeric offspring (*PLEDs*) are not too far behind. Being able to self-assemble supramolecular materials with specific optical and electrical properties will eventually make the synthesis and fabrication of future generations of electronic devices much easier than those prepared using crystalline materials.

Summing Up

Different polymers routinely have different physical properties, obviously. Objects made of cis-polyisoprene are rarely confused with those fabricated of polystyrene. Polymers synthesized from the same monomers can have different properties depending upon how they are synthesized (e.g., cellulose and starch; LDPE [low-density polyethylene] and HDPE [high-density polyethylene]). And the same polymer can be processed in different ways to produce objects with different properties (e.g., polypropylene [PP] carpeting, PP bottles, PP bumpers for automobiles). These three variables—choice of monomer, choice of polymerization method, choice of processing—enable the *enormous* range of properties that is possible with polymers, unlike any other class of materials. The progress of polymer science, just like any other science or technology, depends on a combination of gradual evolution and sporadic revolution.

Many of the objects that we encounter on a daily basis are becoming less passive. Examples include active packaging, changing the way foods are marketed, and smart materials. As electronic devices become even smaller, more sophisticated, and more flexible, we can expect to see an increasing number of consumer products containing sensors that enable the products to respond to some stimulus. This technology will significantly affect medical devices, changing the way drugs are administered and monitored, for example. Microfluidic devices will probably enable the easy sampling and analysis of body fluids, speeding the process of collection, analysis, and diagnosis. Other products will benefit from the use of MEMS devices and electronic imaging.

Nanocomposites are already making an impact on the choice and use of polymeric materials. As the dimensions of the particles diminish into the range of a few nanometers, surface area effects dominate, changing fundamentally the interactions between particle and polymer. Often nanocomposites containing less than 5% additive have substantially improved properties with no adverse effects.

Supramolecular chemistry is chemistry beyond the molecule. It requires the simultaneous, precise control during synthesis of chemical composition, physical properties, and morphology over considerable length scales. An important component is self-assembly, the buildup of molecular structures using nonbonding interactions. We have much to learn from complicated natural polymeric systems, such as proteins, which tend to form one three-dimensional structure in preference to a huge number of other possibilities.

References Cited

Akhras, G. 2000. Smart materials and smart systems for the future. *Canadian Military Journal* Autumn: 25–31.

Alivisatos, A. P. 2001. Less is more in medicine. *Scientific American* September: 67–73.

APME. 2001. *Environmental examples: Material recycling.* Association of Plastics Manufacturers in Europe. *www.apme.org.*

Giannelis, E. P. 2002. Personal communication.

Graff, G., and S. Moore. 1998. O₂ scavengers give 'smart' packaging a new lease on shelf life. *Modern Plastics International* 28(2): 69–72.

Henry, C. 2002. Microfluidic circuits: Lab on a chip. *Chemical and engineering news* September 30: 11.

Knight, J. 2002. Honey, I shrunk the lab. *Nature* 418 August 1: 474–75.

Lahann, J., et al. 2003. A reversibly switching surface. *Science* 299: 371–74.

Lendlein, A., and R. Langer. 2002. Biodegradable, elastic shape-memory polymers for potential biomedical applications. *Science* 296: 1673–76. Supporting material: *www.sciencemag.org/cgi/content/full/1066102/DC1.*

MacGillivray, L. R., and J. L. Atwood. 2001. Spherical molecular assemblies: A class of hosts for the next millennium. In *Chemistry for the 21st century*, eds. E. Keinan and I. Schechter, 131. New York, NY: Wiley-VCH.

Mark, H. 1987. From revolution to evolution. *Journal of Chemical Education* 64 (10): 859–61.

Mestel, R. 2002. Days-old socks that smell as fresh as a daisy. *Los Angeles Times* June 17 S2.

Newnham, R. E., and A. Amin. 1999. Smart systems: Microphones, fish farming, and beyond. *Chemtech* December: 38–47.

Oriakhi, C. O. 2000. Polymer nanocomposition approach to advanced materials. *Journal of Chemical Education* 77 (9): 1138–46.

Sandia National Laboratories. 2003. *http://mems.sandia.gov/scripts/index.asp.*

Sinclair, K. B. 2001. Future trends in polyolefin materials. *Macromolecular symposia* 173: 237–61.

Whitesides, G. M., J. P. Mathias, and C. T. Seto. 1991. Molecular self-assembly and nanochemistry: A chemical strategy for the synthesis of nanostructures. *Science* 254: 1312–19.

Section 4 Demonstrations and Experiments

Demonstrations and Experiments

Introduction

In this section is a collection of activities that can be used as laboratory experiments, demonstrations, or both. Although each unit contains reasonably detailed directions, the reader is encouraged to use these as only a starting point and to engage his or her imagination to further explore the concepts and activities described. Deviations from the suggested procedures need to have the approval of a teacher or other responsible adult completely familiar with any potential hazards.

A note on safety: Some of the activities call for chemicals or reagents that are toxic in some way and/or produce toxic vapors under some conditions. Individuals working with these materials need to understand the types of hazards involved, how to protect themselves and those around them from these hazards, and how to proceed in the event of a spill or accident. Read the entire activity before proceeding! Note the safety precautions and follow them. Chemical safety goggles must be worn at all times. Appropriate chemically resistant rubber gloves must be worn when working with hazardous liquids or solids. Dispose of all reagents and materials in a safe and responsible way. Disposal regulations vary from state to state and locality to locality. Be sure to follow local guidelines. The American Chemical Society has published a safety guide (*http://membership.acs.org/c/ccs/pubs/chemical_safety_manual.pdf*) that teachers should find very useful. Online safety information can be found using SciLinks and the keywords *Safety in the Science Classroom*. As a teacher, if you are uncomfortable allowing your students to work directly with the reagents required for any of these experiments, it would make sense for you to perform that activity as a demonstration.

Free Radical Polymerization

Introduction

This experiment demonstrates bulk polymerization using casting resin. Casting resin is a clear polymer that is used to embed biological specimens, coins, or other objects. It comes as a liquid that typically contains low molar-mass, unsaturated polyester and styrene. The unsaturated polyester contains carbon-carbon double bonds. The polymerization is catalyzed with a "hardener," an organic peroxide dissolved in an organic solvent. The peroxide forms free radicals that initiate polymerization of the styrene and crosslinking of the polyester.

Materials

- quantity of casting resin (available at hobby store or Flinn Scientific)
- bottle of peroxide hardener (available with casting resin)
- shallow containers (e.g., watch glass, evaporating dish, or petri dish) or mold
- tongue depressor or other stirring stick
- object to suspend in the resin (optional)
- forceps or tweezers (if embedding an object)
- small piece of polystyrene (e.g., part of a Tic-Tac box)
- 2 vials or small bottles with caps
- container of acetone
- chemical safety goggles
- chemically resistant rubber gloves
- fume hood or well-ventilated location

Safety

Avoid breathing vapors from casting resin or hardener solution; work in a well-ventilated location. Work carefully to avoid skin contact with any of the liquid reagents; wear chemically resistant rubber gloves. Do not ingest reagents. Avoid eye contact; wear chemical safety goggles. Keep away from heat and open flames. Follow precautions on the containers.

Objectives

In this experiment you will perform a bulk free radical polymerization using readily available starting materials. You will observe the change in physical properties as the reaction proceeds. The polymer object that you make can be either a clear "plastic" piece or a casting containing some small object.

Procedure

Cover the work space with newspaper or drop cloth. Liquid resin or hardener may damage an unprotected surface. Add hardener only to the amount of casting resin that you will use immediately. Once mixed, polymerization will begin quickly. Depending upon the shape of your object, it can have interesting optical properties.

A. **Simple Polymerization.** Mix a small amount of resin and hardener as per the directions. Pour the mixture into a shallow container to form a thin layer, place in a well-ventilated location, and periodically probe the top surface with a tongue depressor or pencil. (While this polymerization is taking place, you can proceed to the next section.) What changes do you observe? When casting is completely hard, remove from mold and examine it. Did the mold that you used have a curved or flat bottom? Hold your casting over a newspaper or textbook. Does it have any interesting optical properties?

B. **Variations.** For this part, it might make sense for students to work in groups, with different students doing separate experiments. Cool a container and a small sample of liquid resin and hardener (separate, not mixed together) in a refrigerator. Remove, mix reagents together, and pour into cold container. Return to refrigerator. Periodically probe top surface as in part A and note any differences. If it is sunny and hot outside, repeat experiment with container, resin, and hardener above ordinary room temperature. Compare the times for the surface to become hardened for all three polymerization temperatures. Add approximately equal parts of casting resin to three small containers. Add one drop of hardener to one sample and two or three drops to the second. Do not add any hardener to the third. Observe as above.

C. **Embedding an Object.** Mix a small amount of resin and hardener as per directions. Fill a mold one-half to one-third full. Allow the polymer to form a gel (several minutes or longer depending upon the thickness). This is the front or top of the casting. Then dip object to be embedded in fresh casting resin and press gently into gel. Fill the mold or container with a second batch of resin and hardener and allow to cure completely. The casting is completely safe to handle when fully cured. A number of small objects can be embedded in a series of layers, creating a three-dimensional effect. See directions that come with the resin for other suggestions. Be creative!

D. **Testing Solubility.** Place a small, thin piece of cured resin in a small vial or bottle. Place a small piece of polystyrene in another vial or bottle. Pour enough acetone into each bottle to cover the samples and then cap. Shake the bottles periodically and/or probe the pieces with a pencil or tongue depressor and note any differences.

Disposal

Cured resin is harmless and can be disposed of with normal trash. Unused liquid resin and hardener should be mixed together and converted to cured resin.

Questions

1. How do you explain the change in physical properties of your casting as it cures?
2. What is the specific role of the peroxide hardener?

3. Was there any effect on the rate of hardening as a function of temperature? As a function of concentration of peroxide hardener? Explain.
4. Is your object optically clear (like window glass), or does it distort light in some way? Explain.
5. Did the acetone affect each piece the same way? Did either dissolve? Explain.
6. What is the difference between a linear and a crosslinked polymer?
7. Why are there safety hazards for the liquid reagents, while the cured solid object is hazard-free?

Discussion

Casting resin ("monomer") with no added peroxide hardener (free radical initiator) does not polymerize at any appreciable rate. Increasing the concentration of hardener should increase the rate of polymerization but not change the properties of the final cured resin. Increasing the temperature should increase the rates of initiation and polymerization. Polystyrene is a linear thermoplastic that dissolves in acetone. The cured casting resin is an extensively crosslinked thermoset material that is insoluble in everything.

References

Chapter 5: "Free Radical Chain Polymerization."
Plastic Embedding Kit. Flinn Scientific, Inc., Catalog No. FB1224.

Interfacial Step-Growth Polymerization: Synthesis of Nylon

Introduction

This experiment is an example of a step-growth polymerization that takes place at the interface of two immiscible solutions. For this reaction, a diamine dissolved in water reacts with a diacid chloride that is dissolved in an organic solvent. Because neither of the monomers is soluble in the solvent containing the other, reaction can occur only at the surface or interface between the two solutions. The product is a polyamide, either nylon-6,6 or nylon-6,10, depending upon the number of carbon atoms in the diacid chloride chosen. This activity works well either as a laboratory experiment or as a demonstration.

Some specialty nylons are actually prepared by the technique of interfacial polymerization. In the laboratory, small quantities (1 to 100 g) can be made in a blender (similar to a kitchen blender, but explosion proof), where the rapid agitation increases the surface area of the mixture, increasing the rate and efficiency of the reaction.

Materials

- few mL of a 0.25 M adipoyl or sebacoyl chloride in cyclohexane or hexane (See "Solution Preparation"1, p. 230.)
- few mL of a 0.5 M hexamethylene diamine in 0.5 M aqueous NaOH (See "Solution Preparation"2, p. 230.)
- acid-base indicator or food dye (optional)
- small beaker or glass jar
- forceps
- stirring rod
- acetone or alcohol (methanol or ethanol)
- larger beaker to wash nylon (optional)
- chemical safety goggles
- chemically resistant rubber gloves
- laboratory apron
- fume hood or well-ventilated location

Safety

Avoid breathing cyclohexane, hexane, adipoyl chloride, or sebacoyl chloride vapors. Work carefully to avoid skin contact with any of the liquid reagents; wear chemically resistant rubber gloves. Do not ingest reagents. Avoid eye contact; wear chemical safety goggles. Cyclohexane, hexane, acetone, and alcohol are flammable; keep away from heat and open flames. Follow precautions on the containers. The nylon "rope" should not be handled with bare hands until after it has been washed thoroughly with alcohol and dried.

Objectives

In this experiment you will carry out a step-growth polymerization that produces a sample of nylon. You will observe the formation of the polymer only at the interface of two immiscible solutions. Basic physical properties such as density and solubility are reviewed. You will gain experience in manipulating reactive chemical reagents safely, isolating product from those reagents, and purifying it. Also, you will learn how to dispose of excess reagents safely.

Reaction (shown with adipoyl chloride)

$$Cl-\overset{\overset{O}{\|}}{C}-(CH_2)_4-\overset{\overset{O}{\|}}{C}-Cl \ + \ H_2N-(CH_2)_6-NH_2 \ \longrightarrow \ \left(\overset{\overset{O}{\|}}{C}-(CH_2)_4-\overset{\overset{O}{\|}}{C}-NH-(CH_2)_6-NH\right)_n \ + \ HCl$$

  adipoyl chloride   hexamethylene diamine   nylon-6,6

Procedure

The amounts of the reagents used depend upon how this experiment is carried out. Individual preparations can be accomplished with 5 to 10 mL of each of the reagents. Demonstrations before entire classes are better performed with 50 to 75 mL of each reagent in a 250 or 400 mL beaker or clear jar. A few drops of indicator solution or food dye can be added to the diamine solution for better visualization.

Wearing chemically resistant rubber gloves, pour enough of the diamine solution into the bottom of the container so that it is no more than $1/4$ to $1/3$ full. Carefully pour an approximately equal volume of the diacid chloride solution down the side of the container so that mixing of the two liquids is minimized. What do you observe? With the forceps, pull the material that forms at the interface of the two solutions straight up and wind it around the stirring rod. Rotate the stirring rod to continue pulling up nylon "rope" and winding it up on the rod.

When you have wound a reasonable quantity of polymer, break the rope with the forceps and submerge the polymer that you have collected on the rod into a beaker of acetone or alcohol. You can unwind the rope with the forceps (do not handle with your bare hands yet) to make better contact with the wash solvent. Leave the polymer in the solvent for at least an hour, or better, overnight. Remove, rinse with a little fresh solvent, then with water, and dry. At this point the nylon should have absolutely no odor and is safe to handle with your bare hands. Examine the properties of a small length of your sample. Is it strong? Does it resemble nylon fishing line?

Disposal

Probably the best procedure for disposing of leftover reagents is to mix all of them together, stir well, and allow the reagents to fully react. Transfer any remaining liquids to a hazardous waste container and dispose of properly. When the solid nylon is dry and has no odor, it can be disposed of in the trash.

Questions

1. Why is the diamine solution added to the container first? What would happen if the adipoyl or sebacoyl chloride solution were poured in first, followed by the diamine solution? What physical property dictates which solution is added first?
2. Why does the nylon form only at the interface of the two solutions?
3. What can you say about the solubility of your nylon sample? Does is dissolve in hexane (or cyclohexane)? Is it water soluble?
4. Why was the rope pulled from the solutions much thicker than the washed and dried polymer? Why were you cautioned not to handle the rope until it had been washed and dried?
5. What are some differences in properties between your sample of nylon and those of monofilament nylon fishing line? How do you explain these differences? (Hint: see Chapter 8, "Fibers.")
6. Drawn below is the structure of sebacoyl chloride. Write an equation for the formation of nylon-6,10.

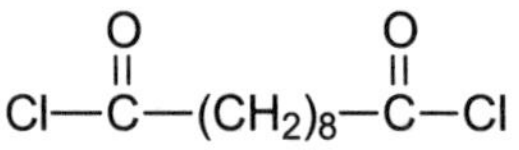

$$Cl-\overset{\overset{\displaystyle O}{\|}}{C}-(CH_2)_8-\overset{\overset{\displaystyle O}{\|}}{C}-Cl$$

Discussion

Nylon forms only at the interface of the two immiscible solutions because neither of the reagents is soluble in the other solvent. In addition, the nylon polymer is not soluble in either solvent. The "rope" pulled from the interface is actually a column of nylon filled with solvent and reagents. After washing and drying, the strand is considerably smaller in diameter. It lacks the strength of something like a monofilament fiber because the molecules in our nylon sample have not been oriented by stretching. In addition, the molar mass of our nylon is probably considerably lower than that of a commercial sample.

References

Chapter 5: "Interfacial polycondensation."
Chapter 7: "Improving Polymer Strength."
Chapter 8: "Fibers."
Mystery Nylon Factory. Flinn Scientific, Inc., Catalog No. AP2088.

Solution Preparation

1. Approximately 0.25 M adipoyl chloride or sebacoyl chloride in cyclohexane or hexane.

Dissolve 4.6 g of adipoyl chloride or 6.0 g of sebacoyl chloride in 100 mL of either cyclohexane or hexane. Store in a capped bottle.

2. Approximately 0.5 M hexamethylene diamine in 0.5 M aqueous NaOH.

Place bottle of hexamethylene diamine in a pan of hot water to melt a small amount of the solid (m. p. 42–45°C). Dissolve 2.0 g of sodium hydroxide pellets in 100 mL of water. Add 5.8 g of hexamethylene diamine and stir. Add a few drops of acid-base indicator or food coloring if desired. Store in a capped bottle.

Step-Growth Polymerization: Synthesis of Polyesters in the Melt

Introduction

In this experiment, both linear and crosslinked polyesters are synthesized by step-growth polymerization. The reactions are carried out in the melt (or *neat*), meaning that no solvents are used in the preparation and the final product does not have to be purified. Preparations such as these are sometimes called "green" reactions.

Materials

- 2 g of phthalic anhydride
- 0.8 mL of dry ethylene glycol
- 0.8 mL of dry glycerol
- 0.2 g of sodium acetate
- acetone
- Bunsen burner, high-temperature laboratory heat gun, or laboratory hot plate
- 2 13 x 100 or 15 x 120 mm test tubes
- 2 small pieces of cardboard
- 2 small stirring rods
- 2 small spatulas
- test tube clamps
- ring stand
- 2 small vials
- Sharpie
- chemical safety goggles
- chemically resistant rubber gloves
- fume hood or well-ventilated location

Safety

Avoid breathing the vapors emitting from the test tubes; work in a well-ventilated area or in a fume hood. These reactions require high temperatures; do not handle hot test tubes with your bare hands, and allow the polymers to cool before handling them. Acetone is flammable. Use in a well-ventilated area and do not allow the liquid to come in contact with your skin. Be sure to wear chemical safety goggles at all times.

Objectives

In this experiment you will carry out two condensation polymerizations that produce samples of distinctly different polyesters, one linear and the other crosslinked. You should appreciate some of the differences in properties of these two classes of materials after performing the experiments below. You should also understand the concept of melt polymerization.

Chemical Reactions Forming Polyesters

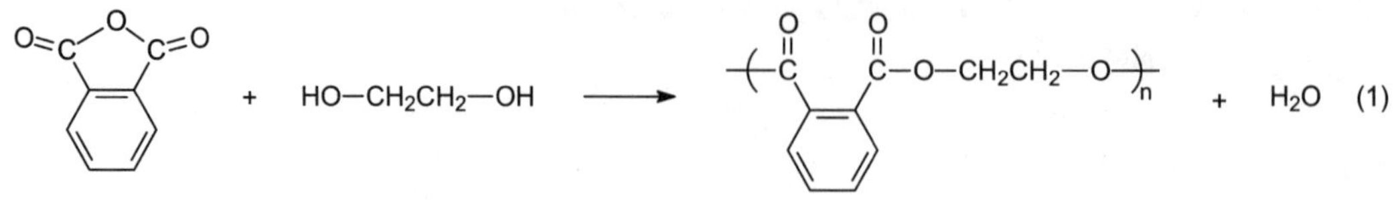

phthalic anhydride   ethylene glycol   poly(ethylene phthalate)

(a diol)   (linear polyester)

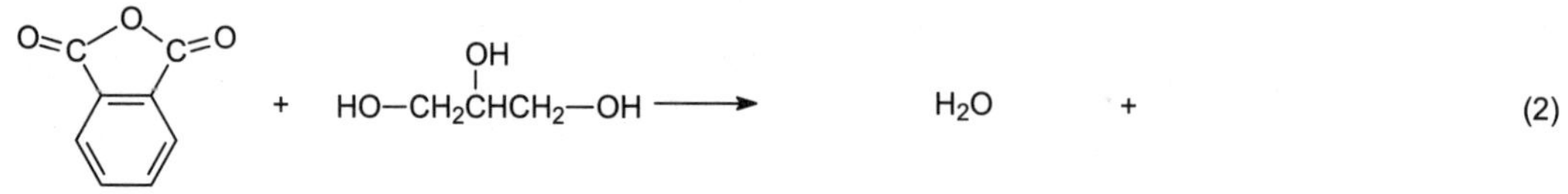

phthalic anhydride   glycerol

(a triol)

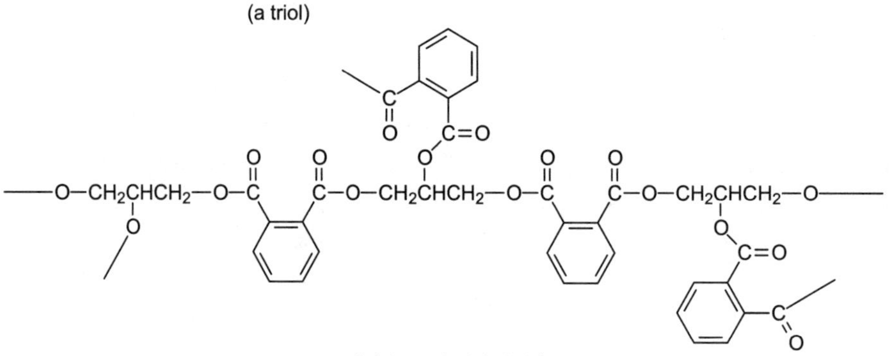

poly(glyceryl phthalate)

(crosslinked polyester)

In equation 1, a dicarboxylic acid derivative called an *anhydride* (literally "without water") reacts with ethylene glycol, a di-alcohol, to form a linear polyester and the byproduct water. Ethylene glycol is the major ingredient in most automotive radiator fluids. In equation 2, the ethylene glycol is replaced with a tri-alcohol, glycerol. As the reaction proceeds, the polyester does not form linear chains, but rather becomes crosslinked as the *three* OH groups react with phthalic anhydride, building up a three-dimensional structure.

Procedure

Place 2 g of phthalic anhydride in each of 2 test tubes labeled #1 and #2. Add 0.1 g of sodium acetate catalyst to each of the tubes. To test tube #1 add 0.8 mL of ethylene glycol and stir. To test tube #2 add 0.8 mL of glycerol and stir. Clamp both tubes in such a way that they can be heated simultaneously. Gently heat the tubes, first to melt the reactants and then to carry out the polymerization. The appearance of bubbles is evidence that the ester is being formed. Continue heating for 4 to 5 minutes, stopping if reaction mixture begins to form a foam. Record your observations for each reaction.

At the end of the heating period, pour the contents of each tube onto a piece of cardboard. Probe each sample with a stirring rod and note any differences. When cool, break off a piece of each sample and transfer to a small, numbered vial. Add a small amount of acetone to each vial, then cap and shake periodically. Does either sample dissolve? Record your observations.

Disposal

Test tubes and samples of polyester can be disposed of the in the trash. Acetone solutions should be placed in an approved disposal container.

Questions

1. Reaction 1 and reaction 2 produce two fundamentally different types of polymer (although both are polyesters). What are these types?
2. What is the vapor that boiled out of the test tubes as you performed the reactions?
3. Did you observe any differences between the reactions in the two test tubes? Describe your observations as the reactions proceeded.
4. Did you detect any differences in the properties of the two products, either as they were poured from the test tubes or after they had cooled? Describe your observations for each.
5. What differences, if any, did you observe in the vials to which you added acetone and polyester?

Discussion

This experiment works well if both the ethylene glycol and glycerol are very dry. Since both are hygroscopic, it is best to take them from sealed bottles. Water contamination causes incomplete polymerization, resulting in product with very low viscosity.

Reaction 1 produces a linear polymer (a thermoplastic) that should be soluble in acetone, while reaction 2 produces crosslinked, insoluble polymer (a thermoset resin). The viscosity of the crosslinked polymer when hot should be noticeably higher than that of reaction 1. Although individual results will depend upon the purity of the starting materials and the heating rate, often the linear product is glassy and hard while the crosslinked one tends to be more brittle and porous. The latter results from the extremely high viscosity that develops as the crosslinked polymer increases in molecular weight.

Reference

Chapter 5: "Making PET in the Melt."

Step-Growth Polymerization: Synthesis of a Polyurethane Foam

Introduction

This experiment is another example of a step-growth polymerization, one that produces a crosslinked polymer. Two liquids are mixed, beginning the chemical reactions that cause polymerization and foam generation. The result is a hard polyurethane foam, similar to the material commonly used for insulation, for flotation in boats and canoes, and in furniture. This activity works well either as a laboratory experiment or as a demonstration.

Materials

- several mL of two-part polyurethane foam system (available from Flinn Scientific; catalog no. C0335)
- food dye (optional)
- at least 3 disposable cups
- 1 latex glove
- 1 600 mL beaker
- Sharpie or other permanent marker that will write on the cups
- small graduated cylinder
- tongue depressors or other disposable stirring sticks
- empty toilet paper or paper towel cylinder
- aluminum foil
- paper towels
- vial or small bottle with cap
- acetone
- chemical safety goggles
- chemically resistant rubber gloves
- laboratory apron
- fume hood or well-ventilated location

Safety

Avoid breathing the vapors of either the liquid starting materials or the curing polymer object; work in a well ventilated area or in a fume hood. Work carefully to avoid skin contact with any of the liquid reagents; wear chemically resistant rubber gloves. Do not ingest reagents. Avoid eye contact; wear chemical safety goggles. Follow precautions that are included with the polyurethane foam system. Polyurethane samples that you make should not be handled with bare hands until after they have hardened and cured. Wash your hands thoroughly on completion of the experiment.

Objectives

In this experiment you will carry out a step-growth polymerization that produces a polyurethane foam. You will observe the formation of the polymer as it also expands in volume. In addition, you should better appreciate the concept of crosslinking following the completion of this activity.

Simplified Reaction for Formation of Polyurethane

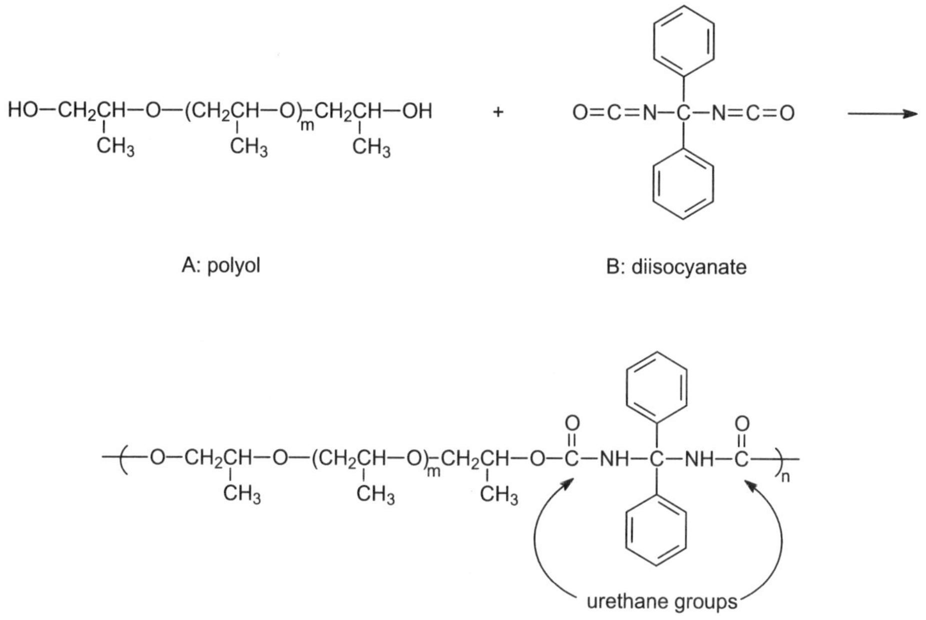

A: polyol

B: diisocyanate

urethane groups

The equation above shows the reaction of a polyether polyol, the major component of part A, with a diisocyanate (in part B), producing a linear polymer. The reagents in your kit include a mixture of starting materials, including oligomers to increase the viscosity of the liquids and trifunctional species that lead to crosslinking. In addition, a foaming agent, surfactant, and catalyst are included. Because the chemistry is complex, only the simplified reaction is shown.

Procedure

Preparing Graduated Mixing Cups. Using the graduated cylinder, measure 10 mL of water into a disposable cup and mark the height of the water with a line on the *outside* of the cup. Add a second 10 mL, and make a second mark. Discard the water and dry out the cup. Prepare a second cup with marks at 30 mL and 60 mL.

A. **Estimating the Volume Increase during Polymerization.** Glue one end of a toilet paper or paper towel roll onto a piece of paper or cardboard so that the roll stands upright on the paper base. Use enough glue so that the bottom of the roll is sealed to the base. Place the cylinder/base on newspaper or paper toweling.

Get your 20 mL graduated cup. Wearing chemically resistant rubber gloves, an apron, and protecting the surface you are working on, pour liquid A directly into your cup up to the 10 mL mark. Pour liquid B on top of liquid A until the total liquid level reaches the 20 mL mark. If you add a little too much, leave it in the cup—do not pour any excess back into the liquid B container! Stir the two liquids well for a few seconds, and then pour the liquid mixture into the upright cylinder. Transfer as much of the mixture from the cup into the cylinder as you can, making sure that the liquid reaches the bottom. Keep liquid off any unprotected surfaces, including your clothing and exposed skin. Observe the reaction mixture as the polymerization proceeds. Guide your polymer with the tongue depressor as necessary to keep it from falling over. After the polyurethane has hardened and is safe to handle, measure the dimensions of your sample and calculate its approximate volume in cm^3. Finally, calculate the number of times the initial volume of the starting liquid increased to form the expanded foam.

B. **Making a Polyurethane Hand.** Place the latex glove inside the 600 mL beaker and pull the opening of the glove over the edge of the beaker so that the glove is suspended inside the beaker with the fingers hanging down. Get your 60 mL graduated cup. Wearing chemically resistant rubber gloves, pour liquid A into the cup up to the 30 mL line, followed by liquid B up to the 60 mL mark. Stir as above and transfer the mixture into the glove. Remove the glove from the beaker and work the liquid so that each finger is filled. Lay the glove on a protected surface and observe the polymerization. As the reaction proceeds, *gently* feel the glove and record your reaction. Do not handle the "hand" until the polyurethane is hard and has cured.

C. **Making Foam Cookies.** Lay out a piece of aluminum foil as a "cookie sheet." Wearing chemically resistant rubber gloves, mix 10 mL of liquid A and 10 mL of liquid B in a graduated cup. Pour the reaction mixture onto your cookie sheet, forming several circles of liquid approximately 6 cm in diameter with at least 4 cm between each circle. Allow to "bake" (cure at room temperature). Do not handle until completely cured, and do not allow anyone to eat any of your cookies.

D. **Making Other Shapes.** Try performing the polyurethane polymerization in other containers. Estimate the amount of reaction mixture you will need so as to produce about the same volume of polymer as the container you choose. Before you proceed, have your plan approved by a teacher or other responsible adult.

E. **Does the Polyurethane Foam Dissolve?** Place a small piece of cured foam from one of your preparations above in a vial or small bottle, cover with acetone, and cap. Periodically shake the mixture over a period of at least 30 minutes and observe. At the discretion of the teacher, try dissolving samples of your polyurethane foam in other solvents as available (e.g., water, ethanol, hexane, toluene).

F. **Examining the Foam.** Break or cut through one of your samples to reveal the inside. Look carefully at the sample and describe the structure that you see.

Disposal

Samples of cured foam can be disposed of in the trash. Acetone solutions should be placed in an approved disposal container.

Questions

1. Why is it necessary to mix samples of liquid A and liquid B together before polymerization proceeds? Could we add a little catalyst to liquid A, say, and achieve the same effect (as we did in the free radical polymerization experiment)?
2. By how much did your cylindrical foam sample expand during polymerization? According to the manufacturer, the liquid expands approximately 30 times when forming the foam. If the number you calculated for your sample is lower than this, suggest some possible reasons for the discrepancy.
3. What happened when you treated a sample of your foam in acetone? If your sample did not dissolve, suggest some reasons why not.
4. Describe what the inside of your foam looks like.

Discussion

This experiment demonstrates a two-part step-growth polymerization. Each part contains monomers and oligomers with functional groups that react with complementary functional groups in the other part. Therefore, unlike chain-growth (e.g., free radical) polymerizations, this step-growth polymerization occurs as soon as both components are mixed. This mixture also contains a foaming agent that produces samples in the form of a hard foam. The manufacturer states that the volume expansion on polymerization is approximately 30. However, the value observed may be less than this and will depend upon the relative amounts of parts A and B that were mixed together. Included in the set are monomers that are tri- or tetra-functional. The polyurethane product therefore is extensively crosslinked, and will not dissolve in any solvent.

References

Chapter 5: "Step-Growth Polymerization."
Chapter 8: "Foamed Objects."
Polyurethane Foam System. Flinn Scientific, Inc., Catalog No. C0335.

Polymer Precipitation

Introduction

Precipitation is a common technique used to isolate and purify polymers. For synthetic polymers, this often involves dissolving the polymer in an organic liquid that is a "good" solvent for the polymer, and then adding this solution slowly to a large excess of a "poor" solvent (or nonsolvent), a liquid in which the polymer is insoluble. Introduction to the poor solvent causes the polymer chains to collapse, aggregate, and come out of solution.

In this experiment, a natural, water-soluble polymer is precipitated into water. Actually, this process is commonly carried out in the kitchen. When the nonsolvent is a certain aqueous solution, the final concoction is called egg-drop soup. A second method for precipitating egg white involves the use of an organic solvent.

Materials
- the whites of two eggs
- beaker or pan of boiling water
- beaker of alcohol (methanol, ethanol, or rubbing alcohol)
- large cooking spoon or spatula
- chemical safety goggles

Safety
Wear chemical safety goggles. Alcohol is flammable; keep away from heat and open flames. Avoid burns from boiling water or steam. If working in the laboratory, do not eat or taste anything.

Objectives

In this experiment we learn about the physical transformation of precipitation. The polymer chosen is egg albumin, a water-soluble protein that becomes insoluble when it is heated or mixed with an organic solvent. We gain experience in manipulating and isolating polymers.

Procedure

A. **Precipitation into Water.** Heat a large beaker or pan of water to boiling. While stirring the boiling water, slowly add the egg white (or beaten whole egg). Allow the water to cool before handling the precipitate.

B. **Precipitation into Alcohol.** Place a quantity of alcohol in a beaker at room temperature, stir, and slowly add the egg white. You can isolate the precipitate by filtration, or by carefully pouring off the alcohol, leaving the polymer behind. Wash it with water before handling.

C. **Attempted Redissolving.** Add a small amount of both of your precipitates to cold water. Do either of the samples redissolve?

Questions

1. What is the major polymer found in egg white? To what class of polymers does it belong?
2. Why is the egg white soluble in cold water, but precipitates in hot water?
3. Why is the precipitated egg white insoluble in cold water?

Discussion

Because the polymer used in this experiment is a protein (albumin), precipitation denatures the polymer, causing permanent changes in its secondary, tertiary, and quaternary structure. Therefore the precipitated polymer cannot redissolve. The denaturing/precipitation occurs because the three-dimensional shape of albumin is sensitive to temperature and to the nature of its environment (stable in water but not in other solvents).

References

Chapter 6: "Random Coils."
Chapter 3: "Proteins."

Gels from Alginic Acid Salts

Introduction

In this experiment you will work with a natural polymer called alginic acid that comes from seaweed. Alginic acid is a polysaccharide as are cellulose and starch. Unlike cellulose and starch, however, alginic acid as the name implies contains acidic functional groups called carboxylic acids. Salts of alginic acid are used as food additives, especially as a thickener.

Polymers containing ions, or functional groups capable of forming ions, can have unusual properties. Alginic acid contains a carboxylic acid group on each repeat unit along the chain. When these groups are in the acid form, they interact with each other strongly through hydrogen bonding, and the polymer is insoluble in water. Converting them to the sodium salt (sodium carboxylate), however, reduces the interactions between the polymer chains, and the ionized polymer (sodium alginate) dissolves in water. This is shown schematically in the following equation:

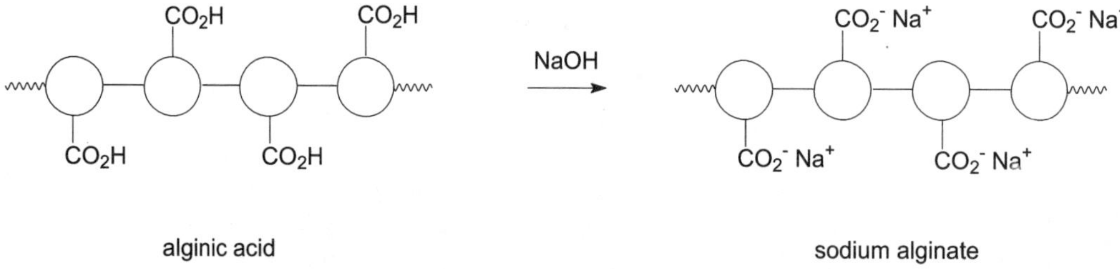

alginic acid → sodium alginate

In the equation, the large circles represent the sugar units that make up the polymer chain, and the squiggly lines indicate that the polymer chain continues in both directions.

While alginate salts with alkali metals such as sodium or potassium are soluble in water, salts with divalent cations such as calcium, copper, and zinc (Ca^{2+}, Cu^{2+}, and Zn^{2+}, respectively) are insoluble in water. In this experiment, you will observe what happens when a drop of a sodium alginate solution is added to an aqueous solution of a divalent cation. The transformation from polymer solution to gel involves the concepts of polymer solubility, diffusion, and ion exchange.

Materials

- 10 mL of 2 mass % solution of sodium alginate in water (See "Solution Preparation"1 on p. 244.)
- 10 mL of 1M HCl solution
- 50-100 mL of 2 mass % aqueous $Ca(NO_3)_2$
- 10 mL of 2 mass % aqueous $CuSO_4$
- 10 mL of 10 mass % aqueous $CuSO_4$
- solutions of other di- or trivalent metal salts as desired

■ saturated NaCl solution
■ concentrated NH_3 solution ("NH_4OH")
■ pH 12 NaOH buffer solution (See "Solution Preparation"2 on p. 244.)
■ distilled water
■ pipets with different tip sizes
■ test tubes or small vials
■ pH test strips
■ small spatula
■ crystallizing dish or beaker
■ chemical safety goggles
■ chemically resistant rubber gloves

Safety

Hydrochloric acid, ammonia and sodium hydroxide solutions are caustic and harmful to skin. Avoid contact with the skin and eyes; wear chemically resistant rubber gloves and chemical safety goggles. Some of the metal salts are toxic. Handle carefully and wash your hands on completion of the experiment or before eating.

Objectives

One of the purposes of this experiment is to demonstrate how polymer chains containing ionic groups interact with each other and with metal ions in solution. Converting the carboxylate ions to free carboxylic acid groups by protonating them diminishes the solubility of the polymer and causes precipitation. Gel formation is observed when drops of polymer solution encounter divalent metal ions, which act as ionic crosslinking agents. The transformation from solution drop to complete gel bead requires an appreciation of ion exchange and diffusion.

Procedure

A. Precipitation. Add 1-2 mL of 2% sodium alginate solution to 10 mL of 1M HCl solution in a test tube or vial and stir or shake. Note your observations.

B. Alginate Gels.
  1. Bead Preparation
     a. Slowly add 10 drops of 2% sodium alginate solution to approximately 10 mL of 2% $Ca(NO_3)_2$ solution. What do you observe? Do you notice any changes over 5 minutes or so? Repeat with 2% solutions of other cation solutions such as $MgSO_4$, $FeSO_4$, $Co(NO_3)_2$, $Ni(NO_3)_2$, $Zn(NO_3)_2$, $Al_2(SO_4)_3$.
     b. Add approximately 10 mL of 2% $CuSO_4$ solution to a small vial or test tube and label it "2% Cu^{2+}." Add approximately 10 mL of 10% $CuSO_4$ solution to a second tube and label it. Add 10 drops of 2% sodium alginate solution first to the 2% and then add 10 drops to the 10% copper sulfate solutions. Compare the beads in these two solutions carefully by holding

the tubes up to the light, shortly after they have formed and then about every 5 minutes for the next 30 minutes or so and write down your observations.

   c. You can make beads of different sizes by adding different sized drops to the divalent cation solutions.

2.   Do the Beads Dissolve?

   a. Repeat the preparation of a few $Ca(NO_3)_2$ beads as described above. Immediately remove one or two of the beads and place them in a tube containing saturated NaCl. Label the tube and set it aside. Periodically observe any changes in the beads.

   b. With a small spatula, remove at least 4-6 beads from the 2% $CuSO_4$ solution and place in a clean vial or test tube. Add distilled water, shake, and carefully pour off the water, leaving the beads in the tube. Repeat with fresh water. This step rinses the beads, removing excess cation. Rinse the spatula.

Prepare a tube containing 10 mL of distilled water and 2 drops of concentrated NH_3 solution. Label the tube "NH_3" and record the pH. Transfer two or three of the rinsed beads to this tube and observe for several minutes. Record your observations.

Prepare a tube containing 10 mL of a NaOH buffer at pH 12.[2] Label the tube "NaOH" and record the pH. Transfer two or three of the rinsed beads to this tube and observe for several minutes. Record your observations. Compare the results of treating the beads with the two bases, NH_3 and NaOH.

C. **Alginate "Snakes".** This makes an excellent demonstration on an overhead projector. Fill a crystallizing dish or beaker to a depth of ½ to ¾ inch with 2% $Ca(NO_3)_2$ solution. Introduce a stream of 2% sodium alginate solution directly into the Ca^{2+} solution using a pipet. You should be able to cause the formation of a long strand of gel (a "snake"). For the most part, the properties of the strands should be the same as those of the beads that you explored above. To test the physical properties of your "snakes," wash them in fresh water and let them dry. As they are drying, periodically test their strength and note their appearance.

Disposal

All solutions can be poured down the drain and followed by water. Beads and "snakes" can be disposed of in the trash.

Questions

1. What caused the sodium alginate to precipitate when it was added to the HCl solution? Write an equation for the reaction that occurred. If you chemically

reversed this reaction, do you think the polymer would dissolve? Describe how you would carry out such an experiment.

2. How did the beads in the 2% $CuSO_4$ solution differ from those that formed in the 10% $CuSO_4$ solution? Are they the same shape? Did they immediately become uniformly cloudy (translucent)? Did they all float on the solutions? What properties changed over time? Explain.

3. You may have observed that beads form at the top of the metal ion solutions and then eventually sink. This indicates that the density of the bead is increasing over time. In other words, the composition inside the bead changes gradually. This was probably most noticeable with the beads that formed in the $CuSO_4$ solutions. Describe the chemical and physical processes that take place from the addition of the initial alginate drop to a divalent metal ion solution to the final, equilibrium bead. (Hint: Think in terms of precipitation, ion exchange, and diffusion.)

4. Did beads form in all of the cation solutions you used? If not, can you explain why some did and some did not?

5. Did the beads that you put in saturated NaCl undergo any changes? How do you explain what happened?

6. Why did the beads placed in ammonia solution behave differently from those treated with NaOH solution? Did you notice any color changes in either solution? What was the significance of the color change? (Hint: Add a few drops of aqueous ammonia to a $CuSO_4$ solution.)

7. Explain why beads formed when drops of alginate solution are added to a divalent metal ion solution, but long strands form when a continuous stream is introduced into the same solution. What would happen if you cut one of the strands into short segments?

8. What physical properties do the dried strands have?

Discussion

Protonating sodium alginate with HCl solution reverses the equation above, causing precipitation of alginic acid. The carboxylic acid groups hydrogen bond extensively with other acid groups, either on the same chain or on different chains. This causes the polymer chains to aggregate and come out of solution. A similar situation occurs when sodium alginate solution is treated with many divalent ions. Ions such as Ca^{2+} complex with more than one carboxylate group, causing ionic crosslinking. The result is an insoluble gel that is swollen with water. In theory it should be possible to redissolve this gel by flooding the system with Na^+ ions. However, practically this does not always happen, particularly with gels that have not been freshly prepared.

Observant students will notice differences in the beads formed over time and in different solutions. Initially, a translucent shell can be observed as the beads float on a given solution. In a few minutes, the entire bead appears translucent, and it will sink in the solution. As a drop of sodium alginate solution is added to a divalent ion solution it is surrounded by, say, Ca^{2+} ions that replace Na^+ on the surface of the drop, forming the insoluble bead. Additional Ca^{2+} ions must first diffuse through this outer shell of calcium alginate before they can replace sodium ions on the interior. As these processes occur, the bead becomes more translucent and its density increases.

Increasing the pH of copper alginate beads with sodium hydroxide will probably not cause the beads to dissolve. Adding beads to concentrated ammonia solution, however, will slowly dissolve the beads because the copper ions complex more strongly with NH_3 (forming the deep blue $Cu(NH_3)_4^{2+}$) than they do with carboxylate ions.

References
Chapter 3: "An Ocean of Polysaccharides."
Chapter 6: "Gels."
Waldman, A. S., et al. 1998. The alginate demonstration: Polymers, food science, and ion exchange. *Journal of Chemical Education* 75 (11): 1430–31.

Solution Preparation
1. 2 mass % sodium alginate
Sodium alginate is available from Flinn Scientific (catalog no. S0445) or Aldrich Chemical Company (catalog no. 18,094-7). For best results, add the sodium alginate to the desired amount of water and allow it to sit or stir overnight. Stirring or shaking the following day will yield a hazy, tan solution. The solution is never absolutely clear.
2. pH 12 NaOH buffer
Combine 25 mL of 0.2 M KCl with 6.0 mL of 0.2 M NaOH and dilute to 100 mL with distilled water.

Identifying Plastics by their Densities

Introduction

In Chapter 7 we pointed out that many polymers have characteristic densities, a physical property that is relatively easy to measure. Thus one can make a good start at identifying an unknown polymer sample by determining its density. Because the densities of most common polymers are known, it is easy to compare the density of the unknown with those listed in a table (for example, in the *Polymer Handbook*; Brandrup, Immergut, and Grulke 1999).

Alternatively, one could use the differences in polymer density for practical purposes. For example, polyolefins such as polyethylene and polypropylene are less dense than water, while poly(methyl methacrylate) and PET are more dense than water. A mixture of plastic pieces of, say, HDPE and PET, could be separated from each other in a recycling operation by flotation of the former on water.

Materials

- several pieces of different kinds of plastic (e.g., HDPE cut from milk jugs, LDPE from lab wash bottles, polypropylene from yogurt cartons, PET from soda bottles, polystyrene from Tic-Tac mint boxes or CD jewel cases, etc.)
- 250 mL beakers, drinking glasses, or small containers
- plastic wrap, aluminum foil, or watch glasses to cover containers
- graduated cylinder
- balance
- forceps
- rubbing alcohol (70% isopropyl alcohol) ("IPA")
- NaCl (table salt)
- vegetable oil (optional)
- glycerine (optional)
- tap water
- large glass cylinder (optional)
- long rod (optional)
- chemical safety goggles

Objectives

In this experiment you will study a common physical property of solids, density. As you discovered in Chapter 7, the density of a polymer is dependent upon its chemical composition, the degree of crystallinity, and how it is processed. Adding fillers to a polymer will produce a composite with a density no doubt different from that of the pure polymer. Therefore, you should appreciate that the density values listed for polymers are approximate. After performing this experiment, you will understand how density is used as a tool to identify polymers. However, you should realize that positive identification of an unknown sample almost always requires additional analytical techniques. You should also appreciate that density can be utilized as a tool in the recycling of commercial polymers.

Procedure

Preparation of Solutions of Known Density. Prepare approximately 200 mL of each of the following liquids or solutions (relative amounts are volumes unless otherwise indicated). For best results, cover the liquids loosely and allow dissolved gases to escape for a day or two before using them.

No.	Solution	Density (g/cm³)
1	IPA/H_2O (85/15)	0.89
2	IPA/H_2O (60/40)	0.93
	or vegetable oil	0.92–0.93
3	H_2O	1.0
4	NaCl in H_2O (10% by mass)	1.1
5	NaCl in H_2O (25% by mass)	1.2
	or glycerine	1.25

A. **Determination of Relative Densities.**

Select a piece of plastic and add it to liquid #1. If it does not sink, push it under the surface with the forceps to make certain that trapped air bubbles are not keeping it afloat. Note whether it is more or less dense than the liquid. If it sinks in liquid #1, test the sample with liquid #2. Repeat until the sample floats in one of the liquids. Write down the approximate density of this sample based on your observations. Similarly test the other samples that you are provided and record your results. If you know the identity of these samples, compare your results with the data in Table 1 on the next page. If you are given an unknown plastic, deduce its possible composition using the data in Table 1.

B. **Variation: Construction of a Gradient-Density Column.**

Instead of using separate containers for each of the test solutions, one can combine them in a tall, narrow column (such as a large graduated cylinder). Using a long rod, carefully pour the liquid with the highest density (liquid #5) down the rod into the bottom of the column. Try not to contact the sides of the column with the liquid. Next, *slowly* pour liquid #4 down the rod, so that the two liquids mix as little as possible. Liquid #4 should be sitting on top of #5. Continue until you have a band of each of the liquids in the column. For best results, loosely cover the column to minimize evaporation and let it sit in a safe place for a few days, allowing dissolved gases to escape. For greater visual effect, add a different color of food dye to each of the liquids before pouring into the column. The solutions should not be too dark, however.

Now, add a piece of plastic to the column and see where it comes to rest. If you know its identity, you should be able to predict how far down the column it will travel. This makes an effective demonstration.

246

Questions

1. Why is it recommended that dissolved gases be removed from the test liquids before using them?
2. Explain how the shape of a plastic sample might cause an erroneous density measurement.
3. How would you measure the density of a plastic bottle?
4. What are some of the reasons why the density of a polymer can vary from sample to sample?
5. Rigid plastic pipe and flexible laboratory tubing (e.g., Tygon tubing) are both made of poly(vinyl chloride). The pipe has a density of 1.4 g/cm^3, while the tubing has a density of only about 1.2. Explain how they can be so different.

Table 1. Densities of common plastics.

Plastic	Density (g/cm^3)
polypropylene (PP)	0.85 – 0.92
low density polyethylene (LDPE)	0.89 – 0.93
high density polyethylene (HDPE)	0.94 – 0.98
acrylonitrile-butadiene-styrene terpolymer (ABS)	1.04 – 1.06
polystyrene (PS)	1.04 – 1.08
nylon-6,6	1.13 – 1.16
poly(methyl methacrylate) (PMMA)	1.16 – 1.20
polycarbonate (PC)	1.20 – 1.22
poly(ethylene terephthalate) (PET)	1.38 – 1.41
rigid poly(vinyl chloride) (PVC)	1.38 – 1.41
poly(tetrafluoroethylene) (PTFE)	2.1 – 2.3

Discussion

Pouring or mixing solutions incorporates a small volume of air in the liquid. Over time, these gases will diffuse out of the liquid and form bubbles on some surface. Trying to perform this experiment when gas bubbles are forming on the sample pieces can be very frustrating and will provide erroneous results. For best results, the samples used in this experiment should be pure polymer, not samples containing plasticizer or fillers. Poly(vinyl chloride) samples are almost always plasticized. Rigid PVC parts usually contain a filler.

References
Chapter 7: *"Density."*

Brandrup, J., E. H. Immergut, and E. A. Grulke, eds. 1999. *Polymer handbook*, 4th ed. New York, NY: John Wiley and Sons.

Bruzan, R., and D. Baker. 1993. Plastic density determination by titration. *Journal of Chemical Education* 70: 397–98.

Kolb, K. E., and D. K. Kolb. 1991. Method for separating or identifying plastics. *Journal of Chemical Education* 68: 348–49.

Woodward, L. 2002. *Polymers all around you*, 2nd ed., 11. Middletown, OH: Terrific Science Press.

Yoon, R.-H. 1997. Methods of separating used plastics for recycling. *Chawong Risaikring* 6 (2): 113–31.

Experiments with Films

Introduction

Many plastics are fabricated into thin films. Films find widespread applications, including the backing for photographic and X-ray "film," self-cling kitchen wrap, and plastic bags. An average HDPE grocery bag has a mass of less than 7 g, approximately $1/8$ the mass of a typical paper grocery sack, and is approximately only 0.02 mm thick, also about $1/8$ the thickness of its paper cousin. However, it can safely hold an incredible load. In this experiment we will explore a property of polymer films called cold drawing, working with a variety of common films. *Cold drawing* refers to the room-temperature stretching or deformation of a polymer sample.

Materials

- plastic dry cleaning bag (LDPE; recycling code 4)
- clear plastic kitchen wrap
- plastic lunch or freezer bags
- plastic grocery bag (HDPE; recycling code 2)
- ruler
- scissors
- paper cutter (optional)
- fine permanent marker

Objectives

We will use readily available plastic films to demonstrate stress-strain behavior. Students should be able to relate the physical behavior of thin films to the concepts of orientation and crystallinity. They should be able to explain terms such as cold drawing, yielding, and machine and transverse directions.

Procedure

Hold a piece of newspaper, notebook paper, or paper toweling near one corner, and tear down slowly. Hold the piece of paper near another corner, and tear in a direction at right angles to the original tear. You should notice that you were able to tear in one direction a reasonably straight, narrow strip of paper parallel to the edge of the sheet. This is called the machine direction (MD). In the other direction, however, you probably were unable to make a straight tear. This is called the transverse direction (TD). Although with the naked eye paper does not appear to have any structure, the cellulose fibers that make up most inexpensive paper tend to line up more completely in the direction in which the paper was manufactured (thus the term "machine direction"). Tearing a sheet of paper in this direction produces reasonably straight edges because the tear runs along the length of these oriented fibers. Tearing in the transverse direction requires primarily breaking or pulling apart fibers that tend to be oriented at roughly right angles to the direction of the tear.

We can think of paper as being a crude analog of polymer films. The molecules in films tend to become oriented during manufacture, more so in the machine direction than in the transverse direction. Hold a piece of polymer film up to the light. If a series of small lines is visible, the machine direction is parallel to those lines. If no lines are evident, arbitrarily label one direction "A" and the right-angle direction "B."

Cut a few strips of film approximately 1cm by about 5 cm in the machine direction (or "A" direction), and label one end of each (MD or "A"). The exact dimensions are not important. However, the sides should be parallel and have smooth edges. A sharp paper cutter might work well for cutting the strips. Similarly, cut a few strips in the transverse direction and label them (TD or "B"). In the center of two of the pieces (one MD or A, the other TD or B), mark off a 2 cm section. Keep these strips for part B. See example below.

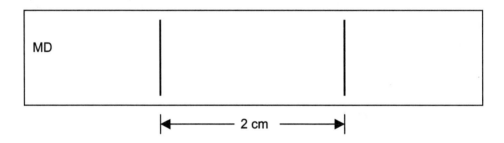

A. **Qualitative Tests.** Hold an unmarked TD strip at each end. Quickly pull the strip apart and observe the behavior. Take another strip, and this time start to pull the strip slowly, then release. What do you observe? Repeat the pulling and releasing, but each time pull a little harder. Finally, slowly pull the strip, keeping force on it without releasing it. Watch various parts of the strip as you stretch it. It might be instructive to hold each strip up to the light and look through it as you stretch them. Repeat these operations with strips cut in the machine direction. Record your observations and note whether the MD and TD strips exhibited the same behavior.

B. **Quantitative Tests.** Take a piece of film cut in the transverse direction and having the 2 cm marks on it. Measure the width of these pieces in between the 2 cm marks. Holding one of the strips up to the light, pull gently and evenly until the marked section is uniform and the lines are gone. While still applying pressure, measure the new length of strip between the marks and record your result. Allow the strip to relax for 5 minutes. Remeasure the length between the marks and the new width. Record. Perform this test on a strip cut in the machine direction.

C. **Tearing Samples.** Holding a new TD strip at one end, try to tear it in half down its length. Repeat, but use the thin part of the TD strip that you stretched in part B. Repeat with two MD strips. Record your observations.

D. **Other Film Samples.** Repeat these tests with other samples of film that are available. Record your observations and compare results with different kinds of films.

Questions

1. Did the samples from your bags cut in the machine direction behave the same as those cut in the transverse direction? How might you explain any differences?

2. If your bag had no visible lines, can you determine the machine direction based on the stretching behavior of films labeled "A" and "B"?

3. Did any of your samples undergo a change in optical properties while being stretched (e.g., become cloudy or opaque)? Can you suggest a reason for this behavior?

4. As you pulled on a piece of film, did it suddenly become much easier to stretch it? What is the phenomenon called when the film suddenly elongates with less force?

5. In part B, by how much did your TD film elongate when measured under stress? After you let the sample relax for 5 minutes, how much longer was the stretched region compared to the original 2 cm? What were your results for the MD sample?

6. Did any of your samples tear easily along their length? Which ones? How do you explain such behavior?

Discussion

You should notice that when you pull on a strip slowly, the film stretches at first, but then recovers. This is an example of *elastic* behavior. On further pulling the film is permanently deformed and does not completely recover. This is mainly *viscous* behavior. At some point the film may "neck" and "yield." *Necking* refers to the narrowing down of a sample on being elongated. Essentially you are observing the rate at which polymer molecules move. If you stretch a piece of film quickly, it breaks because the molecules cannot move fast enough to reorient. Pulling more slowly allows the long, entangled molecules to move past each other and to stretch out. As elongation continues, you should reach the *yield point*, the point at which the sample is no longer elastic and begins to flow (or undergo cold drawing). Often as the sample yields, elongation requires much less force. As the molecules elongate, they tend to form crystallites with adjacent molecules, increasing the strength of the sample.

Molecules in the machine direction are more completely aligned than those in the transverse direction. The film has already been stretched and therefore cannot elongate much further. In the TD, however, there is considerably more disorder. Stretching causes molecules to align and form crystallites. Depending upon the sample of film that you used, the elongation in this direction can be substantial, 3 or 4 times the original length. As the film elongates, the thickness and width of the sample decreases accordingly.

References

Chapter 7: "Stress-Strain Properties" and Figure 7-3.

Chapter 7: "Orientation."

Chapter 8: "Cast and Blown Films."

Spencer, R. D. 1984. The dependence of strength in plastics upon polymer chain length and chain orientation. *Journal of Chemical Education* 61 (6): 555–63.

Sources of Additional Experiments and Demonstrations

Forming Crosslinked Gels

Slime, Ooze...

This gel is usually prepared by adding 1 part of a 4 mass % aqueous sodium borate (borax) solution to 3 or 4 parts of a 4 mass % aqueous poly(vinyl alcohol) solution.

Some Sources of Materials

Slime demonstration kit (among other products)   Flinn AP1829
Fluorescent slime   Flinn AP9081

References

Anon. 2002. Make slime. *Flinn Chemical and Biological Catalog*. 441.

Casassa, E. Z., A. M. Sarquis, and C. H. Van Dyke. 1986. The gelation of poly(vinyl alcohol) with borax. *Journal of Chemical Education* 63 (1): 57-60.

Maynard, B. 2001. It's time for slime. *ACS Chemistry* (Spring): 10-12.

Stroebel, G. G., J. A. Whitesell, and R. M. Kriegel. 1993. Slime and poly(vinyl alcohol) fibers: An improved method. *Journal of Chemical Education* 70 (11): 893.

Gluep

The polymer for this gel is common white glue (mostly poly[vinyl acetate]), crosslinked as above with borax.

Reference

Anon. 1998. What's Gluep? Characterizing a cross-linked polymer. JCE classroom activity #11. *Journal of Chemical Education* 75 (11): 1432B.

Gak

Adding talcum powder to the preparation above makes a very elastic gel.

Reference

Anon. 2002. Make Gak. *Flinn Chemical and Biological Catalog*. 439.

Sodium Polyacrylate (Superabsorbent Polymer)

Pour 300 to 400 mL of deionized or distilled water into a beaker containing 1 g of sodium polyacrylate, stir, and obtain a solid gel in just a few seconds.

A Source of Materials

Sodium polyacrylate   Flinn W0012

Ghost Crystals
A jar apparently containing only liquid actually is filled with invisible "crystals."

A Source of Materials

Ghost crystals   Flinn G0050

Non-Newtonian Fluids

Poly(ethylene oxide) Solution
Solutions of this high molar mass, water-soluble polymer have very unusual properties.

Some Sources of Materials

Super-duper polymer gel	Flinn AP4556
Poly-ox with a twist	Flinn AP5931
Superliquid	Dynamic Development Co. 6801

References

Super Liquid. Flinn Scientific, Inc., Publication No. 1923.
Super Duper Polymer. Flinn Scientific, Inc., Publication No. 4556.

Polymers with a "Memory"

Some semicrystalline polymer samples are stretched above their glass transition temperature and then cooled, freezing in the stress in the new shape, which is often thinner and stronger than that of the original. Examples include shrink tubing, heat-shrink film, PET soda bottles, and some food container lids (e.g., Dannon yogurt). Reheating these objects causes them to relax to their original shape. Heating can be done with a heat gun, or by placing the empty and dry object in an oven (usually ~325°F or ~160°C is sufficient). Note the decrease in area and increase in thickness as most objects shrink.

Some Sources of Materials

Shrinky Dinks, Frosted	Flinn AP1966
Shrinky Dinks, Printable	Flinn AP5451

Reference

Shrinky Dinks. Flinn Scientific, Inc., Publication No. 1966.

Commercial Suppliers

Flinn Scientific, Inc.	Dynamic Development Company
P. O. Box 219	P. O. Box 582
Batavia, IL 60510	El Toro, CA 92630
800-542-1261	
www.flinnsci.com	

Glossary

Glossary

Acrylic—Refers to monomers or polymers of acrylic acid (CH_2=$CHCO_2H$) and its derivatives. Poly(butyl acrylate), poly(methyl methacrylate), polyacrylamide, and polyacrylonitrile are acrylic polymers.

Addition polymer—A polymer prepared from an addition reaction, almost always to a monomer with one or more carbon-carbon double bonds. Polyethylene, poly(methyl methacrylate), and polystyrene are addition polymers.

Adhesive—A substance capable of holding materials together by surface attachment through forces such as dipole-dipole interactions.

Amorphous polymer—A polymer with random chain arrangements or conformations; a polymer can be entirely amorphous or it can contain some crystalline regions.

Anionic polymerization—A polymerization in which the initiator and intermediate species are anions.

Antioxidant—A compound added to a polymer, normally in small amounts, to prevent degradation by reactions with oxygen or ozone.

Aromatic polymer—A polymer with a predominance of benzene ring-containing repeat units. Polystyrene is an example.

Atactic polymer—A polymer in which the groups bonded to the main chain are arranged in random spatial orientations. A polymer with an absence of tacticity.

Biodegradable—A polymer that can be broken down by biological organisms such as enzymes, bacteria, fungi, or algae.

Block copolymer—A copolymer with long sequences or blocks of different repeat units (e.g., A_xB_y).

Blow molding—A process used in the fabrication of plastic bottles and other hollow shapes; a heated tube of plastic (a preform or parison) is forced against the sides of a mold by air pressure.

Blowing agent—A chemical added to a monomer or polymer system to generate a gas that becomes trapped inside the polymer matrix, resulting in a foam.

Branched polymer—A polymer in which the main chain contains some number of side chains or branches.

Brittleness—A measure of the ease with which a polymer breaks when an attempt is made to deform it.

Cationic polymerization—A polymerization in which the initiator and intermediate species are cations.

Chain-growth polymerization—Polymerization in which a monomer reacts with an initiator, forming a reactive species that propagates the molecular chain until it ends in a termination step.

Commodity resins—High-volume, low-price thermoplastics (polyethylene, polypropylene, poly[vinyl chloride], polystyrene and other styrenics, and poly[ethylene terephthalate]).

Composite—A material that contains two or more structurally different components with properties different from that of any individual component. Examples include crosslinked polyester resin reinforced with glass fiber and rubber filled with carbon black.

Compounding—Processing operation in which fillers, plasticizers, coloring agents, and other additives are mixed with a polymer, often in an extruder, to obtain desired properties.

Compression molding—A process in which polymeric objects are formed in a mold by the application of heat and pressure.

Condensation reaction—A reaction in which two dissimilar molecules react, forming a new molecule and a small-molecule byproduct such as water.

Coordination polymerization—A polymerization in which propagation occurs when monomer and a growing polymer chain both coordinate to a transition metal catalyst and then react. The final polymer often consists of one particular tacticity, e.g., isotactic polypropylene.

Copolymer—A polymer that contains more than one type of monomer.

Crosslinked polymer—A three-dimensional polymer created when intermolecular forces connect adjacent chains; the forces may be hydrogen bonds, dipole interactions, van der Waals forces, or ionic or covalent bonds.

Crystalline polymer—A polymer with ordered chain structure that allows the formation of crystallites.

Crystallite—A small region within a polymer in which the chains are packed together, forming a regular crystal structure.

Curing—A chemical reaction in which crosslinking occurs to produce a network polymer.

Degradation—1. The decrease in degree of polymerization of polymer chains. 2. The negative change in physical properties or appearance of a polymer.

Degree of polymerization—Average number of monomer units in a polymer molecule. Abbreviated DP.

Dendrimer—A highly branched, tree-like macromolecule with ordered structure and usually globular shape.

Dimer—A molecule consisting of two monomer units bonded together.

Elastic elongation—The reversible change in length of a material when it is stretched and the stretching force is then removed.

Elastomer—A rubbery polymer; is easily deformed under low stress and recovers its original dimensions when the stress is removed (e.g., a rubber band).

Emulsion polymerization—Free-radical polymerization carried out in micelles suspended in water. Ingredients include a surfactant (detergent) to form the micelles, a monomer that is not very soluble in water, and an initiator. The product consists of small particles of polymer suspended in water called a "latex."

Extrusion—The processing step in which a threaded shaft (a screw) melts a polymer sample and forces it through a die that produces the desired shape.

Fiber—A natural or synthetic crystalline polymer that exists in long, thin strands.

Filler—A relatively inexpensive substance added to a polymer during processing to add bulk and perhaps to improve properties. (e.g., carbon black in rubber used for automobile tires).

Film—A polymer processed into a very thin sheet.

Foam—A substance that is a blend of a polymer and a gas. Polystyrene packing peanuts and polyurethane cushions or insulation are common examples.

Free-radical (or radical) polymerization—A chain-growth polymerization in which the initiator and intermediate species are free radicals (species with at least one unpaired electron).

Gel—A chemically or physically cross-linked polymer that is highly swollen with solvent. Called a hydrogel if the solvent is water (e.g., gelatin).

Glass transition temperature—The temperature at which an amorphous polymer (or the amorphous fraction of a semicrystalline polymer) changes from a hard glassy state to a rubbery, flexible state; abbreviated T_g.

Graft copolymer—A branched copolymer in which chains made up of one type of repeat unit are bonded to a chain made up of a different repeat unit.

Homopolymer—A polymer comprised of a single kind of monomer or repeat unit.

Hydrophilic—A compound that is attracted to water (water-loving).

Hydrophopic—A compound that repels water (water-hating).

Initiator—A chemical substance that helps bring about a polymerization reaction; unlike a catalyst, an initiator is changed by the reaction and may appear in the final product.

Injection molding—A process used in the fabrication of thermoplastics; a melted polymer is injected, under pressure from an extruder, into a steel mold and then allowed to cool and solidify.

Interfacial polymerization—A polymerization that takes place at the interface of two immiscible liquids. Usually it is a step-growth polymerization in which one monomer is soluble in one of the liquids, and the other monomer is soluble in the other liquid (e.g., nylon "rope").

Isotactic polymer—A polymer in which like groups bonded to the main chain are always on the same side of the chain; isotactic polymers have a high degree of crystallinity (e.g., isotactic polypropylene).

Latex—1. The milky liquid obtained from certain plants and trees, including rubber trees. 2. The product of an emulsion polymerization.

Linear polymer—A polymer in which monomer units are linked in long, one-dimensional macromolecules (long chains).

Living polymerization—A polymerization in which the reactive intermediates have very long lifetimes. An example is an anionic polymerization with no termination steps.

Macromolecule—A polymer molecule is a macromolecule.

Modulus—The ratio of stress (force) to strain (elongation) for a given polymer.

Molar mass—The average mass of a polymer molecule equals DP times molar mass of the monomer.

Monomer—The simplest unit of a polymer. A chemical species that can be converted to a polymer.

Morphology—The study of the form and structure of a substance.

Non-Newtonian fluid—A substance whose viscosity changes depending upon the force applied to it. Examples include mayonnaise, latex paint, and Silly Putty.

Orientation—The straightening and alignment of polymer chains when a sample is stretched. This often leads to an increase in crystallinity.

Parison—A hollow, injection-molded tube that when blowmolded becomes a bottle.

Photodegradable—A polymer that can be broken down when exposed to light.

Plastic—Usually refers to a thermoplastic polymer (as distinguished from an elastomer or a thermoset polymer).

Plasticizer—An additive that softens a thermoplastic by reducing the glass-transition temperature or reducing crystallinity.

Polymer—A large molecule (a macromolecule) made up of many (poly) repeat units (mers). A sample of macromolecules.

Polymerization—A chemical reaction that can be used to convert monomers into polymers.

Preform—See Parison.

Random coil—The thermodynamically favorable shape a polymer chain assumes in the absence of applied forces such as stretching.

Random copolymer—A copolymer that has a random sequence of two or more monomer units.

Reactive intermediates—Very reactive chemical species such as anions, free radicals, or cations that form as starting materials are converted to products in a chemical reaction.

Repeat unit—A chemical unit (e.g., a monomer unit) bonded to two other like units, forming the "links" in the polymer chain.

Resin—Usually an amorphous thermoplastic.

Ring-opening polymerization—The process in which a cyclic monomer yields polymer when the ring opens and forms an intermediate that reacts with another cyclic molecule.

Shear—A force caused by one plane sliding past another (e.g., a pair of shears).

Silicone—Polysiloxane, a polymer with a backbone consisting of alternating silicon and oxygen atoms.

Spinning—The extrusion of a polymer through a die containing many small holes (a spinneret), producing fibers.

Stabilizer—A substance added to a polymer to inhibit the degrading effects of oxidizers, ultraviolet light, or electrical discharge.

Step-growth polymerization—Polymerization that proceeds by discrete, stepwise reactions between dissimilar di-

functional compounds (monomers and larger intermediates).

Stiffness—Often referred to in polymer technology simply as modulus.

Strain—Elongation. Percent change in length of an object under stress.

Stress—The force acting across a unit area of a material.

Syndiotactic polymer—A polymer in which like groups bonded to the main chain alternate regularly from one side of the chain to the other.

Tacticity—Describes the stereoisomers that result from the relative positions of groups along a polymeric chain of carbon atoms. Three simple patterns are atactic, isotactic, and syndiotactic.

Tensile strength—The resistance to deformation or breakage of a polymer sample when a pulling force is applied.

Thermoforming—A fabrication process in which sheets of thermoplastic material are softened by heating and then shaped by pressing into a mold.

Thermoplastic—A polymeric material that repeatedly softens and melts when heated and resolidifies upon cooling.

Thermoset—A polymeric material that melts with initial heating, but solidifies permanently with further heating because of extensive crosslinking.

Toughness—The property that describes the stress-strain relationship of a polymer. The area under the stress-strain curve is a measure of toughness.

Trimer—A molecule consisting of three monomer units bonded together.

Vinyl—1. Loosely a monomer with the structure $CH_2=CHR$ or $CH_2=C(R_1)R_2$ (e.g., $CH_2=CHCl$ is vinyl chloride). 2. A polymer prepared from a vinyl monomer (e.g., poly[vinyl chloride]).

Viscoelastic—The property describing a material that behaves either like an elastic solid or a viscous liquid, depending upon the force placed upon it (e.g., Silly Putty).

Viscosity—The resistance of a substance to flow.

Vulcanization—The crosslinking of rubber.

Yield point—The point at which the elastic limit of a material is exceeded; above this point, the material deforms or breaks.

Index

Index

*Page numbers printed in **boldface** type refer to tables or figures.*

National Science Teachers Association